NATIONAL ACADEMIES *Sciences* *Engineering* *Medicine*

NATIONAL ACADEMIES PRESS
Washington, DC

Impacts of National Science Foundation Engineering Research Support on Society

Committee on Extraordinary Engineering Impacts on Society

Program Office

National Academy of Engineering

Consensus Study Report

NATIONAL ACADEMIES PRESS 500 Fifth Street, NW Washington, DC 20001

This activity was supported by contracts between the National Academy of Sciences and the National Science Foundation (EFMA-2101725). Additional support was provided by the National Academy of Engineering's President's Initiative Fund. Any opinions, findings, conclusions, or recommendations expressed in this publication do not necessarily reflect the views of any organization or agency that provided support for the project.

International Standard Book Number-13: 978-0-309-72295-7
International Standard Book Number-10: 0-309-72295-0
Digital Object Identifier: https://doi.org/10.17226/27873
Library of Congress Control Number: 2024945519

Cover images:
Left – Vertical-axis wind turbines. Caltech Field Laboratory for Optimized Wind Energy. California Institute of Technology. Used with permission.
Top right – The Utah Bionic Leg. University of Utah College of Engineering. Used with permission.
Bottom right – 3D printing process. Alexey Bubryak/Getty Images

This publication is available from the National Academies Press, 500 Fifth Street, NW, Keck 360, Washington, DC 20001; (800) 624-6242; http://www.nap.edu.

Suggested citation: National Academies of Sciences, Engineering, and Medicine. 2024. *Impacts of National Science Foundation Engineering Research Support on Society.* Washington, DC: National Academies Press. https://doi.org/10.17226/27873.

COMMITTEE ON EXTRAORDINARY ENGINEERING IMPACTS ON SOCIETY

DAN ARVIZU (Chair), Former Chancellor/President, New Mexico State University System; Former Director, Department of Energy's National Renewable Laboratory, Former Chair of the National Science Board, currently member of the President's Council of Advisors on Science and Technology

EDWARD H. FRANK, Co-Founder and Chief Executive Officer, Brilliant Lime, Inc. and Cloud Parity

SELDA GUNSEL, President, Shell Global Solutions (U.S.); Vice President of Global Lubricants and Fuels Technology, Shell

WILLIAM S. HAMMACK, William H. and Janet G. Lycan Professor of Chemical and Biomolecular Engineering, University of Illinois Urbana-Champaign

EBONEY HEARN, Executive Director, Office of Engineering Outreach Programs, Massachusetts Institute of Technology

LAURA A. LINDENFELD, Dean, School of Communication and Journalism, Vice Provost for Academic Strategy and Planning, Executive Director of the Alan Alda Center for Communicating Science, Stony Brook University

THERESA A. MALDONADO, Vice President for Research & Innovation, University of California Office of the President

LOUIS A. MARTIN-VEGA, Former Dean of Engineering, Distinguished University Professor of Industrial and Systems Engineering, North Carolina State University

YU TAO, Associate Professor of Sociology, Stevens Institute of Technology

JIMMY WILLIAMS JR., Senior Vice President and Chief Technology Officer, ATI, Inc.

JEFFREY R. YOST, Director, Charles Babbage Institute for Computing, Information, and Culture; Research Professor, History of Science, Technology and Medicine, University of Minnesota, Minneapolis

National Academy of Engineering staff
DAVID A. BUTLER, J. Herbert Hollomon Scholar, Study Director
COURTNEY HILL, Program Officer (until January 2023)
CASEY GIBSON, Associate Program Officer (from January 2023)
MAIYA SPELL, Senior Program Assistant (until January 2023)
CHESSIE BRIGGS, Senior Program Assistant (from March 2023)
SUDHIR SHENOY, Associate Program Officer (from December 2023)
GURU MADHAVAN, Norman R. Augustine Senior Scholar, Senior Director of Programs

Contract Writer
STEVE OLSON, Freelance Writer

Reviewers

This Consensus Study Report was reviewed in draft form by individuals chosen for their diverse perspectives and technical expertise. The purpose of this independent review is to provide candid and critical comments that will assist the National Academies of Sciences, Engineering, and Medicine in making each published report as sound as possible and to ensure that it meets the institutional standards for quality, objectivity, evidence, and responsiveness to the study charge. The review comments and draft manuscript remain confidential to protect the integrity of the deliberative process.

We thank the following individuals for their review of this report:

DAVID D. CLARK (NAE), Massachusetts Institute of Technology
ELIZABETH DICKEY, North Carolina State University
NICHOLAS M. DONOFRIO (NAE), NMD Consulting, LLC
PETER GALISON, Harvard University
MARIA GINI, University of Minnesota
HE (HELEN) HUANG, University of North Carolina and North Carolina State University
SUBRAMANIAN S. IYER, National Institute of Standards and Technology
KRISTINA M. JOHNSON (NAE), Columbia University
K.J. RAY LIU (NAE), Origin AI
ANDRE W. MARSHALL, George Mason University
KISHOR C. MEHTA (NAE), Texas Tech University
RICHARD M. MURRAY (NAE), California Institute of Technology
MELISSA ORME (NAE), The Boeing Company
WILLIE PEARSON, Georgia Institute of Technology
JAMES D. PLUMMER (NAE), Stanford University
DONNA RILEY, The University of New Mexico
DIETRAM A. SCHEUFELE, University of Wisconsin-Madison
EUGENE SPAFFORD, Purdue University
ROBERT F. SPROULL (NAE), University of Massachusetts at Amherst
ALAN I. TAUB (NAE), University of Michigan
JEAN W. TOM (NAE), University of Virginia
KON-WELL WANG, University of Michigan

Although the reviewers listed above provided many constructive comments and suggestions, they were not asked to endorse the conclusions or recommendations of this report nor did they see the final draft before its release. The review of this report was overseen by **Anita K. Jones** (NAE), University of Virginia and **Asad M. Madni** (NAE), University of California, Los Angeles. They responsible for making certain that an independent examination of this report was carried out in accordance with the standards of the National Academies and that all review comments were carefully considered. Responsibility for the final content rests entirely with the authoring committee and the National Academies.

Preface

As the chair of the National Academies of Sciences, Engineering, and Medicine committee on Extraordinary Engineering Impacts on Society, it is my profound honor to introduce this report, which showcases the transformative impact of engineering achievements on society, enabled by the National Science Foundation's (NSF's) steadfast support of research and education. This document is a testament to the enduring power of engineering innovation to reshape our world, enhance our quality of life, and drive forward the frontiers of human knowledge and capability.

Engineering is the invisible hand that molds our future, crafting solutions to complex challenges and expanding the horizons of what is possible. From the internet to clean energy technologies and beyond, engineering innovations have fundamentally changed how we live, work, and connect with one another. These achievements, however, did not materialize in a vacuum. They are the fruits of a deeply rooted commitment to research and education, championed by the NSF since its inception in 1950.

This report is not merely a chronicle of past successes; it is a beacon, illuminating the path for future generations of engineers. It underscores the critical role of federal support in fostering an environment where bold ideas can thrive, interdisciplinary collaborations flourish, and educational opportunities abound. By highlighting the stories of individuals whose lives and work have been shaped by NSF-funded projects, we aim to demystify the field of engineering, bringing to light the human ingenuity and perseverance behind each breakthrough.

The contributions of the committee members and the National Academies staff who authored this report cannot be overstated. Their dedication, expertise, and collaborative spirit were instrumental in bringing this complex narrative to life. Together, they have woven a rich tapestry of discovery that not only honors the legacy of engineering innovation but also points the way toward a future brimming with potential.

Central to this report is our intent to inspire the next generation of scientists and engineers, particularly those from backgrounds historically underrepresented in these fields. It is our firm belief that the richness of diversity is our nation's greatest strength, providing the broad range of perspectives and ideas necessary to tackle the societal challenges and aspirations of our time. By showcasing the extraordinary achievements enabled by NSF research and education funding, we hope to kindle a passion for engineering in young minds from all corners of society, encouraging them to envision themselves as the architects of tomorrow's innovations.

In closing, let us reflect on the boundless possibilities that lie at the intersection of engineering and societal need. May this report serve not only as a record of what has been accomplished but also as a clarion call to those who will build the future. The journey ahead is filled with challenges, but it is through embracing these challenges that we forge a better world. To the aspiring engineers and scientists reading this: your vision, your creativity, and your dedication are the keys to unlocking the next great era of engineering innovation. The future is in your hands.

Dan Arvizu, *Chair*
Committee on Extraordinary Engineering Impacts
on Society

Contents

Boxes, Figures, and Tables

BOXES

FIGURES

TABLES

Summary

Engineering continually reshapes society, influencing almost every aspect of our daily lives. It is the innovative force behind such marvels as the internet, bionic limbs, three-dimensional (3D) printing, clean-energy technologies such as wind turbines, and earthquake-resistant buildings. Engineering advances economic growth, enhances manufacturing capacity, enhances infrastructure resilience, improves healthcare, and strengthens national security.

However, the impact of engineering research—and, by extension, engineering education—is often hidden from the public eye. Promoting comprehension of how engineering affects society is thus a crucial aspect of garnering and sustaining public backing of policies aimed at ensuring that technology continues to serve the greater good of humanity.

One part of this effort is highlighting the pivotal role of federal support of engineering research. Such support furnishes researchers with the necessary resources to pursue ambitious projects, fostering exploration into fundamental questions, groundbreaking technology and engineered systems developments, and practical applications. Free of the need to generate profits, government agencies can prioritize research that yields societal benefits, stimulating innovation, promoting interdisciplinary collaboration, and cultivating a skilled workforce.

Since its inception in 1950, the National Science Foundation (NSF)—an independent federal agency—has played a critical role in funding cutting-edge research, including in engineering. In keeping with this responsibility, it charged the National Academies of Sciences, Engineering, and Medicine (the National Academies) to explore how its support of engineering research and education has led to positive societal impacts, focusing on the stories of the people responsible for these impacts. As part of this effort, the committee formed to undertake this work was asked to develop clear, compelling narratives for the public about the sources and effects of engineering innovations and offer recommendations on how to bring this information to the attention of diverse audiences.

STRUCTURE AND ARRANGEMENT OF THE REPORT

Chapter 1 of the report elaborates on the motivations for the study. It presents the statement of task for the committee convened to carry out the work and the insights on fulfilling the task gained from the sponsor, explains the committee's approach to fulfilling the task, discusses where this report fits within the scholarship addressing the impacts of engineering research support, and explains how the committee defined "engineering" and "impact" in the context of its work. The history of NSF's funding of engineering research is explored in Chapter 2, including the units involved in supporting engineering over time and the mechanisms the agency has for providing support. Chapter 3 outlines the many ways in which engineering research and education affect society, the role that the NSF has played in bringing these impacts about, and the means that it employs to evaluate societal impacts. The committee's process for identifying exemplary engineering impacts made possible by NSF investments is presented in

Chapter 4, along with descriptions of those impacts. The final chapter of the report, Chapter 5, addresses the means for communicating engineering's societal impacts and the audiences for that messaging. It summarizes the example outreach materials developed under the committee's supervision and provides guidance on how to reach and engage diverse audiences with information on engineering's impact on society.

Report appendices contain the agenda for the committee's information-gathering symposium, thorough descriptions of the example outreach materials, and biographies of the committee and staff responsible for the work.

REPORT SYNOPSIS

This report is intended to draw attention to engineering's impacts on society, highlighting the stories of people who have benefited from NSF research and education support and who have brought about technological innovations and new types of engineered systems as a result of it. As its statement of task notes, such impacts might take the form of expanded technological and social capabilities, scientific breakthroughs, and improvements in economic opportunity. They could have led to improvements in individual quality of life, national security, population health, manufacturing services, infrastructure resilience, and public policy, among others.

The emphasis on stories and people distinguishes this effort from many other reports produced by the National Academies, which take a more analytical approach to evaluating a particular topic. These include reports that quantify the impact of federally funded research support; examine the conduct and result of various NSF program initiatives; offer advice on communicating information on science, technology, and the value of supporting research to the general public; and raise public awareness of engineering and the role of engineers. Unlike nearly all of these publications, this report was designed for a wide readership.

The content of the report is summarized below.

NSF's Support of Engineering-Related Research and Education

NSF has funded research on engineering and the education of engineers since its founding. These activities were initially undertaken by the Division of Mathematical, Physical, and Engineering Sciences, which focused on fundamental research questions relevant to multiple disciplines rather than awards in single fields. Funding expanded in the late 1950s when scientific achievements in the Soviet Union spurred policies aimed at accelerating U.S. technological capacity, and in 1964, NSF created a separate Division of Engineering with a specific mandate to support engineering research and education. This division increased the funding of graduate students and early career faculty as well as continuing previous initiatives fostering interdisciplinary collaborations.

At the beginning of the 1970s, the agency instituted the Interdisciplinary Research Relevant to the Problems of our Society program. Significantly, this program required that proposals explicitly discuss potential societal impacts. The effort was later absorbed into the Research Applied to National Needs (RANN) program, which addressed several issues with substantial social components, including alternative energy sources and pollution detection.

The Directorate for Engineering was established in 1981. Its mission was to strengthen the technology programs of the agency while maintaining close connections between engineering research and education and related activities in the sciences. The broad institutional structure established at its founding remains largely intact today. As of early 2024, the directorate—now one of eight within the agency—has five major units (NSF, 2024b):

1. Division of Chemical, Bioengineering, Environmental and Transport Systems
2. Division of Civil, Mechanical and Manufacturing Innovation
3. Division of Electrical, Communications and Cyber Systems
4. Division of Engineering Education and Centers
5. Office of Emerging Frontiers and Multidisciplinary Activities

Many other parts of NSF support engineering research and education, both directly and indirectly, as well. Three have a particularly strong focus on the discipline. The Directorate for Computer and Information Science and Engineering, established as an independent administrative unit in 1986, has the mission "to enable the U.S. to uphold its leadership in computing, communications, and information science and engineering" (NSF, 2024a). That same year, the Directorate for Science and Engineering Education—now called the Directorate for STEM Education—was established. Its outreach extends to all educational levels and settings, with a particular focus on broadening the participation of groups that have been underrepresented in engineering. And in 2022 the Directorate for Technology, Innovation and Partnerships came into existence with the goal of collaborating with the other directorates to accelerate the translation of research results to the marketplace and society while cultivating new educational pathways that lead to a diverse and skilled future technical workforce.

NSF has used a wide variety of mechanisms to support such work. The most prominent and visible of these are the Engineering Research Centers (ERCs). The concept of centers was introduced in 1984 as a means to conduct multidisciplinary, systems-oriented engineering research on problems critical to industry and has since been widely adopted elsewhere in government. ERCs have evolved since their inception from a focus on interdisciplinary research at a single host university with industry engagement, through expansions involving multiple partner universities, strategic planning, diversity initiatives, and outreach to pre-college educational institutions as well as inclusive partnerships emphasizing convergent research and workforce development. As of early 2024, NSF had supported 79 ERCs, with more slated for later in the year. It was estimated in 2022 that the centers had collectively yielded over $75 billion in new products and processes on an investment of less than $2 billion (NSF, 2022a; p. 3).

Among many other support mechanisms are Industry-University Cooperative Research Centers (IUCRCs), where academic researchers connect with industry partners to create and sustain collaborations that bridge the gap between the two; the Innovation Corps (I-Corps), a program that facilitates faculty and students in taking the initial steps in commercializing their discoveries with the help of industry mentors, and the NSF Small Business Innovation Research program (SBIR), where small businesses undertake translational research to develop prototypes and scale up production. The agency has also long recognized the importance of developing early career faculty via such initiatives as the Presidential Young Investigator Award, CAREER Award, and Presidential Early Career Award for Scientists and Engineering programs.

In these ways, NSF's engineering programs have had a widespread influence on society through their support of work that brings about new technologies and economic advancement through public–private collaborations and the development of a well-educated workforce.

Considerations in Identifying Engineering Impacts on Society

The committee examined the many ways in which engineering research and education impact society along with NSF's role in bringing these impacts about as part of developing their approach to identifying exemplary engineering impacts made possible by NSF investments. Fundamentally, it is clear that engineering has profoundly shaped society over time through innovations like electrification, the automobile, and fiber optics, and it will have a major role in providing solutions to future challenges such as economical solar energy, universal access to clean water, and enhanced virtual reality. A set of National Academy of Engineering (NAE) reports identified areas of societal import where engineering has or will drive change, with *The Greatest Engineering Achievements of the 20th Century* (Constable and Somerville, 2003) addressing the former and the "Grand Challenges for Engineering" (NAE, 2017) the latter.

Research underpins the important engineering breakthroughs mentioned in these reports, and research requires sources of support. While commercial interests fund and conduct most research and development work in the United States, the federal government still plays an important role. NSF is among the major sources of federal dollars, and it has been particularly influential in lending support to technologies such as the internet in the period between their initial development and the time at which they become commercially viable. Furthermore, NSF programs like the Engineering Research Centers foster interdisciplinary research and academic–industry partnerships that might not otherwise have come to pass. The agency's Materials Research Science and Engineering Centers, Small Business Innovation Research program, and the National Nanotechnology Initiative are just three examples of the means by which impacts are being generated.

NSF's research support programs do not operate in isolation, however. Indeed, complementary federal agency support has been instrumental in driving engineering innovation in the United States. This complementary support may take different forms. In some circumstances, differences in mission drive when a particular agency is likely to be the primary support mechanism. The three federal advanced research project agencies, for example, are tasked with funding cutting-edge, high-risk research efforts that may, if they show promise, be fostered through their next stage of development by other agencies like NSF. In other cases, particular agencies maintain infrastructure like laboratories and testing facilities that other entities can utilize, saving costs and allowing better utilization of specialized operations and support staff. And research in areas that overlap agency responsibilities may be supported by multiple agencies, each contributing to and deriving value from the piece of that work that is consonant with their mission. Overall, the complementary support from federal agencies has been a cornerstone of engineering innovation in the U.S., enabling groundbreaking research, fostering cross-sector collaborations, and providing essential resources that promote continuous advancement in engineering.

NSF is unique, though, among federal funding sources in requiring program and project proposals to delineate their "broader impacts" on society. This core component is defined as the potential of the research to benefit society and contribute to the achievement of specific desired societal outcomes (NSF, 2019). Since its introduction in 1997, the broader impacts criterion has addressed an array of topics from workforce development, national security, and economic competitiveness to science, technology, engineering, and mathematics (STEM) education and diversity equity, and inclusion.

Scholars have analyzed how broader impacts are categorized, with teaching and training being commonly cited in applications, along with dissemination, infrastructure enhancement, and public scientific literacy and engagement. Cultural and political factors influence broader impact considerations, suggesting a need for improved understanding and communication of societal benefits. While NSF's efforts to create institutional capacity around the criterion have progressed, the literature suggests that gaps remain, requiring a better assessment of whom benefits from broader impacts and the establishment of evaluation metrics.

Recognizing Engineering Impacts on Society Brought About by NSF Investments

NSF has itself undertaken three efforts over recent years to identify and document the effects of its support for research and education on the economy, human well-being, and societal advancement.

Developed for the 50th anniversary of NSF in 2000, the *Nifty 50* (NSF, 2000) lists innovations and discoveries catalyzed by agency funding that have become commonplace in daily life. Nearly half of the list, which covers the breadth of the NSF's support initiatives, cite inventions, technologies, or programs related to engineering, including bar codes, Doppler radar, fiber optics, MRI (magnetic resonance imaging), and volcanic eruption detection. Ten years later, the *Sensational 60* (NSF, 2010) list added such engineering advances as biofuels and clean energy, deep sea drilling, and supercomputer facilities. The 70th anniversary contribution was a mural called the "History Wall" (NSF, 2022b), which depicted still more innovations like carbon nanotubes and search-and-rescue robots. These efforts highlighted the substantial return on investment of government support for science, engineering, and technology, and they served as resources to inform the committee's considerations of engineering impacts on society.

The committee undertook two outreach efforts intended to expand upon this information, their own knowledge of engineering innovations brought about by NSF investments, and the background research conducted by staff. One of these was an information-gathering symposium carried out in fulfillment of the statement of task. The other was a set of questionnaires circulated to members of the NAE, along with a companion solicitation of input from NSF staff.

The symposium, which took place in August 2022, touched on major themes raised in the statement of task. It addressed NSF's role in fostering engineering innovations and the people, programs, and funding processes responsible for those innovations. The objectives of the symposium were three-fold: to highlight the rich interconnections between science and engineering and the many ways in which each informs the other, to emphasize the deep and productive link between research and education, and to demonstrate that an underappreciation or misunderstanding of engineering is partially responsible for hindering both technological progress and the participation of underrepresented groups that could contribute to engineering advancements as well as gain benefits from their participation. *Extraordinary Engineering Impacts on Society: Proceedings of a Symposium* (NASEM, 2023) documents the event in detail.

The committee also circulated two questionnaires to members of the NAE. These distinguished educators, researchers, and public and private sector leaders were invited to respond to a set of questions including "Do you have knowledge of any significant engineering impacts on society resulting from funding provided by the National Science Foundation?" Input was solicited from NSF staff on the same question. Their responses spanned the disciplines of aerospace, biomedical, chemical, civil, computer, electrical, environmental, industrial, manufacturing, and mechanical engineering along with computer science, operational systems, and education and workforce development, illustrating the broad reach of the agency. Computer science and engineering garnered the greatest number of responses, with NSF's support of the people and research underlying the technologies that led to Google receiving multiple mentions.

The committee's decision-making process for identifying engineering impacts made possible by NSF investments was guided by this body of information. Primary among their own considerations was that they would not single out a "top 10" but would instead highlight *exemplary* impacts showcasing the breadth and magnitude of NSF's influence, including a variety of engineering disciplines, forms of impact, types of support provided, and recipients. In keeping with the statement of task directive to "engage young people from all segments of society," it featured stories intended to engage and inspire diverse audiences and placed special emphasis on highlighting the work of researchers from historically underrepresented groups in engineering. Recognizing that these impacts are rarely the work of one agency alone, credit is given to other agencies and funders who also contributed to the achievements.

The 10 exemplary engineering impacts on society brought about by NSF investments that were identified by the committee run the gamut from specific technologies to areas of research to programs that provide support. They are briefly summarized below.

1. **Additive manufacturing**, notably 3D printing, has transformed traditional manufacturing with improved material use, design adaptability, and faster production.
2. **Artificial intelligence** models are demonstrating an expanding repertoire of abilities, including text manipulation, pattern recognition, image recognition, self-navigating vehicles, and increasingly accurate speech recognition.
3. **Biomedical engineering** has significantly contributed to society by developing technologies that improve healthcare outcomes and, in the field of rehabilitation engineering, the daily lives of people with disabilities.
4. **Cybersecurity** research leads to technologies, tools, and training that protect against online threats to individual privacy, essential services, and national security.
5. NSF has long been a major player in **engineering education and early career development** through their work with researchers, faculty, and institutions, addressing systemic and structural barriers and promoting equitable opportunities, thereby shaping a more diverse and productive STEM workforce.
6. Many advances in fields as diverse as construction, drug delivery, computing, telecommunications, and transportation are due to research in **materials science and engineering**.
7. The **Engineering Research Centers (ERC) program** has led to groundbreaking research and educational initiatives and demonstrated new types of engineered systems in multiple cross-disciplinary fields that address national challenges, foster innovation, and yield significant gains for the U.S. economy.
8. **NSF contributions to internet advancements** such as NSFNET, TCP/IP and DSL and the agency's early support of such trailblazing researchers as Google founders Larry Page and Sergey Brin have changed the world.
9. **Semiconductors and integrated circuits**—the "brains" behind the electronic devices that we rely on for nearly every aspect of our lives—are possible because of tools like the Metal Oxide Semiconductor Implementation Service (MOSIS) and technologies like fin field-effect transistors (FinFET).
10. Pioneering work in **wind energy technology** includes the design of a vertical-axis wind farm inspired by the formations of schooling fish, which has the goal of significantly enhancing wind farm power density.

Communicating Engineering Impacts on Society to Diverse Audiences

As already noted, any effort to convey the impacts that engineering has had on society encounters an immediate problem: most Americans, including most students and their teachers, have very little exposure to what engineers do or how their work affects everyday life. Increasing support for engineering research and education can yield several positive outcomes, such as enhanced funding for innovative projects, an influx of skilled individuals into the field to meet workforce demand, and informed decisions in the marketplace, leading to high-quality goods and services. In a world facing complex problems that demand creative and technologically sophisticated responses, engineering must be at the forefront of developing solutions.

Research on communicating the impacts of engineering is relatively scarce, but the work that has been done provided the committee with insights. Studies conducted by the NAE in the early 2000s developed and tested messages that were successful in conveying positive perceptions of the discipline to diverse audiences of students and adults. Later research drawing on principles derived from the field of advertising examined the application of concepts like branding and framing to the issue. Storytelling was found to be a particularly effective means of communicating about engineering, as people tend to respond to narratives more strongly than they respond to facts, and narratives are easier to comprehend than traditional scientific explanations.

Studies indicate that messages about engineering, to be successful, must also be tailored to appeal to specific audiences. Hands-on activities are well suited to K–12 students, for example, and they can stimulate interest in engineering's practical problem-solving mindset as well as encouraging relevant skills such as systems thinking, creativity, and collaboration. Women, African Americans, Hispanics, American Indians, people with disabilities, and members of other minority groups remain significantly underrepresented in engineering in the United States. Given that research shows that messages are more effective when they are delivered by people who belong to the same underrepresented group as an audience, it is important to develop messages and encourage communicators who can appeal to these groups.

As the U.S. population becomes increasingly diverse, engineering needs to account for the values and concerns of all people. This objective can be achieved most effectively by having an engineering workforce that reflects that diversity, which means reaching the members of underrepresented groups with the messages, information, and assistance they need to help them join the field.

Previous work has tackled the issue of communicating across purposes and audiences using different labels and mechanisms, including framing, branding, narrative, and storytelling. These share the central idea of curating and presenting information in ways that makes connections to specific audiences, belief systems, mental models, and the like. This report focuses on narratives and related constructs, while acknowledging the complexities of these overlapping areas of scholarship and practice.

The statement of task directed the committee to "[p]rovide guidance on how to reach and engage diverse audiences . . .; promote better understanding of the vital role of engineering in government, business, and society; and engage young people from all segments of society to encourage pursuing a career in engineering." To fulfill this task, the committee contracted with the Alan Alda Center for Communicating Science and with the Massachusetts Institute of Technology to develop a set of example materials for NSF to consider in outreach efforts. Five examples were developed, aimed at different segments of the general public, with an emphasis on students and on diverse groups that may otherwise be poorly informed about engineering and engineers. They use a variety of media (short videos, interactive graphics, workbooks, etc.) and take different approaches (inspirational, educational, humorous, and so on) to engage their audiences. Their content is summarized in Table S-1.

TABLE S-1 Summary of Example Engineering Impacts Outreach Materials Intended for Diverse, General Population Audiences

Title	Intended audience[s]	Media Format	Content Summary
Meet an Engineer	Young people (high school) in historically marginalized groups in STEM (especially underrepresented gender/racial/ethnic groups)	Video-sharing social media platform video (2–5 minutes long)	Interview with a high-profile, NSF-funded engineer from an underrepresented group
Queen of Carbon "Family Tree" (inspired by the work of Millie Dresselhaus)	High schoolers (especially young women)	Clickable interactive image-based web post	Graphic of the many forms of carbon connected in a "family tree"-style diagram illustrating their real-world applications
Earthquake Shake Table	General public (teens and up), especially those who live in high-hazard earthquake areas	Video-sharing social media platform short-form video (2 minutes)	Video illustrating how structures are endangered by earthquakes and how shake tables can be used to test designs
Grand Challenges in Engineering	Elementary and middle school students	Fill-in-the-blank style workbook for elementary students; blog posts for middle school students	Workbook with prompts encouraging students to identify a problem and then draw a device that solves the problem; blog posts highlighting stories of inspiring engineers
Extraordinary Impacts of Engineering	Gen Z/younger millennials	Video-sharing social media platform short-form video (2 minutes)	Animated video illustrating how the technology of an everyday device like a phone has advanced

REPORT CONCLUSIONS AND RECOMMENDATIONS

As the committee's research and the narratives it developed to support its list of exemplary engineering impacts make clear, NSF funding of engineering research and education has had profound societal effects. Based on this and the additional information presented, the committee came to these conclusions:

- NSF's investments in engineering education and research have played a catalytic role in advancing the science, technology, and engineering ecosystems.
- NSF's support of interdisciplinary and intersectoral collaboration on research initiatives and of centers has contributed to engineering's positive impacts on society.
- NSF investments in women and in others underrepresented in the engineering field and in fostering a more supportive learning and research environment for these groups have had a part in bringing about significant engineering contributions.

Additionally, the statement of task requested "conclusions and recommendations on how to best promote understanding of engineering's place in society and how NSF contributes to it." The committee's review of the literature addressing the communication of scientific and technical information to the public leads it to conclude that:

- There is an opportunity to change the way that engineers and engineering are perceived by the general public by highlighting the many ways that engineering affects everyday life, the contributions that engineers make to improving those lives, and the role that investments in engineering education and research play in making these contributions possible.
- Outreach efforts that are grounded in the research on engagement and communication are more likely to reach target audiences and have the impact they intend. Research indicates that efforts that use or highlight people with whom or content with which target audiences can relate are more effective.

These materials also form the basis for the following recommendations regarding the communication of engineering impacts on society:

NSF, in their outreach efforts regarding the support of engineering education and research, should

- **draw upon the literature and experts on public engagement and communication to better target their messaging.**
- **increase the participation and diversity of organizations, people, and voices who have not been well represented in the engineering profession in their messaging.**
- **employ communication forms (short-form videos and media that can be consumed on phones, for example) and forums (including social media platforms) that are used by their target audiences.**
- **feature diverse and relatable people and stories that illustrate how engineering is making everyday life better and how engineers improve the lives of others.**
- **incorporate tracking of the effectiveness of specific messaging efforts into outreach efforts.**

REFERENCES

Constable, G., and B. Somerville. 2003. *A century of innovation: Twenty engineering achievements that transformed our lives.* Washington, DC: Joseph Henry Press. https://doi.org/10.17226/10726 (accessed February 22, 2024).

NAE (National Academy of Engineering). 2017. *NAE grand challenges for engineering.* Washington, DC: The National Academies Press. https://nae.edu/187212/NAE-Grand-Challenges-for-Engineering. (Note that this report was originally published in 2008 and updated in 2017.)

NASEM (National Academies of Sciences, Engineering, and Medicine). 2023. *Extraordinary engineering impacts on society: Proceedings of a symposium.* Washington, DC: The National Academies Press. https://doi.org/10.17226/26847.

NSF (National Science Foundation). 2000. *Nifty 50.* https://www.nsf.gov/about/history/nifty50/ (accessed February 22, 2024).

NSF. 2010. *NSF sensational 60.* https://www.nsf.gov/about/history/sensational60.pdf (accessed February 19, 2024).

NSF. 2019. *Proposal & award policies and procedures guide.* https://www.nsf.gov/pubs/policydocs/pappg19_1/index.jsp (accessed February 22, 2024).

NSF. 2022a. *FY 2020 engineering research centers program report.* https://www.nsf.gov/pubs/2022/nsf22104/nsf22104.pdf (accessed March 12, 2024).

NSF. 2022b. *NSF history wall.* https://www.nsf.gov/about/history/historywall/NSF_history-guide_July2022_Update.pdf (accessed February 19, 2024).

NSF. 2024a. *Directorate for Computer and Information Science and Engineering (CISE).* https://nsfpolicyoutreach.com/policy-topics-/directorate-computer-information-science-engineering-cise (accessed March 12, 2024).

NSF. 2024b. *Engineering (ENG).* https://www.nsf.gov/dir/index.jsp?org=ENG (accessed March 12, 2024).

1
Introduction

This chapter provides basic information about the motivation of this report and the conduct of the study. It presents the committee's statement of task along with insights gained from the sponsor. The committee's approach to fulfilling its task is then detailed, followed by brief descriptions of the conduct of the study and its information-gathering activities, an explanation of where this report fits within National Academies of Sciences, Engineering, and Medicine scholarship addressing the impacts of engineering research, and the definitions that the committee applied for "engineering" and "impact" in the context of its work. The chapter concludes with a summary of the report's organization.

MOTIVATIONS FOR THE STUDY

Despite the pervasive influence of engineering on virtually every aspect of daily life, the public often underestimates or overlooks its profound impact. There are numerous reasons for this lack of awareness. Prominently, it is a consequence of the disconnect that exists between the complexities of engineering processes and their visible outcomes. Most people interact with the products of engineering—be it smart phones, bridges, controlled-release fertilizer, or countless others—without fully grasping the sophisticated research, design, and innovation behind them. Moreover, the interdisciplinary nature of engineering complicates public comprehension, as the engineering process may involve the application of knowledge from such diverse fields such as physics, biology, mathematics, and computer science. As a result of these and other factors, a significant portion of the public fails to recognize how engineering contributes to economic growth, societal advancements, and improved quality of life.

Fostering understanding of the impacts of engineering on society is essential for establishing and maintaining public support for the policies needed to ensure that technology continues to benefit humanity. Federal support of engineering research plays a critical role in this process. Government funding provides researchers with the resources they need to pursue ambitious, high-risk, high-return projects that might not otherwise be feasible. This support enables exploration into fundamental scientific questions, the development of groundbreaking technologies, and the translation of research findings into real-world applications. Crucially, federal agencies often prioritize research that has the potential to generate significant societal benefits, such as improving public health, strengthening national security, or promoting general economic growth. By investing in engineering research, governments can stimulate innovation, foster collaboration across disciplines, and cultivate a highly skilled workforce capable of tackling complex problems on a global scale. Ultimately, federal support of engineering research is essential for advancing knowledge, driving economic competitiveness, and enhancing the well-being of individuals globally.

Since its establishment in 1950, the National Science Foundation (NSF) has played a significant role in supporting engineering research and education in the United States. Initially focused on basic scientific research, NSF expanded its mandate to include engineering in the late 1950s, recognizing the discipline's importance in driving technological innovation and economic growth. Over the years, NSF has funded a wide range of projects across a number of engineering disciplines. Its support has enabled researchers to make significant advances in areas such as infrastructure development, renewable energy, telecommunications, and healthcare technology. Through grants, fellowships, and research centers, the agency has helped cultivate a vibrant engineering research community, fostering collaboration among academia, industry, and government agencies. Today, NSF continues to be a major supporter of engineering research, investing in cutting-edge projects that address pressing societal challenges and push the boundaries of scientific knowledge.

THE COMMITTEE'S STATEMENT OF TASK
AND INSIGHTS GAINED FROM THE SPONSOR

Against this backdrop, NSF asked the National Academies to form an expert committee to undertake a study illuminating how the practice of engineering and the work of engineers has affected society and, particularly, the impacts of NSF support of engineering research. An expert committee was formed to respond to that request.

Box 1-1 contains the statement of task for that committee.

Box 1-1
The Committee's Statement of Task

The National Science Foundation has requested that the National Academies provide it with help in its efforts to bring greater understanding of and attention to engineering's role in fulfilling NSF's mission "to promote the progress of science; to advance the national health, prosperity and welfare; [and] to secure the national defense... ." To achieve this, an ad hoc committee will:

- **Identify up to 10 extraordinary engineering impacts made possible by NSF investments** in research from 1950 onward. These impacts might include expanded technological and social capabilities, scientific breakthroughs, and improvements in economic opportunity. They could have led to improvements in individual quality of life, national security, population health, manufacturing services, infrastructure resilience, and public policy, among others.

- **Organize a virtual public symposium** that highlights how NSF investments in engineering education, research, careers, and institutions propels their mission to "advance the national health, prosperity, and welfare; and to secure the national defense," and produce a proceedings that summarizes the event's presentations.

- **Develop clear, compelling narratives for public engagement** in these engineering impacts, including specific contributions of NSF in fostering these developments.

- **Provide guidance on how to reach and engage diverse audiences with these narratives**; promote better understanding of the vital role of engineering in government, business, and society; and engage young people from all segments of society to encourage pursuing a career in engineering.

The resulting consensus report will be designed for a wide readership, will expand on the proceedings, and offer conclusions and recommendations on how to best promote understanding of engineering's place in society and how NSF contributes to it.

[**Emphasis** added]

These tasks were further elaborated by Dr. Susan Margulies, NSF's Assistant Director for Engineering, in her September 2021 charge to the committee (Margulies, 2021). Dr. Margulies noted that NSF's merit review process relies on two major criteria, both of which address the impact of the proposal. The Intellectual Merit criterion "encompasses the potential to advance knowledge," while the Broader Impacts criterion[1] "encompasses the potential to benefit society and contribute to the achievement of specific, desired societal outcomes" (NSF, 2023). Margulies said that the stories and people underlying these impacts are compelling and that drawing attention to them is a powerful means of fostering greater understanding of and attention to engineering's role in fulfilling the agency's mission. Impact stories, she said, illustrate how fundamental research and innovative education modalities translate into societal benefits; propel prospective visions and new research directions; embolden creativity and risk; and, most importantly, inspire, motivate, and connect. The National Academies study should therefore have as its goals to uncover and illustrate how fundamental engineering research since the origin of NSF has led to positive impacts on American lives, to lend understanding of how the agency's investment in fundamental engineering research contributed to these impacts, and to develop

---

[1] The Broader Impacts criterion is discussed in further detail in Chapter 3.

clear, captivating narratives for the public about recent engineering innovations that improve our lives.

THE COMMITTEE'S APPROACH TO FULFILLING ITS TASK

The committee undertook a wide-ranging information gathering effort to inform its responses to the elements of the statement of task. It surveyed the literature on a variety of topics relevant to the issues and questions raised by the tasks, conducted a symposium to gather information and highlight some engineering achievements, circulated questionnaires to members of the National Academy of Engineering and solicited input from NSF staff. The Alan Alda Center for Communicating Science was contracted to develop draft outreach materials intended to engage diverse audiences in engineering through stories of technology innovations and the people responsible for them.

This work is set forth below and in subsequent chapters of the report and its appendices.

WHERE THIS REPORT FITS WITHIN THE SCHOLARSHIP ADDRESSING THE IMPACTS OF ENGINEERING RESEARCH SUPPORT

National Academies committees are often charged with assessing the state of an area of research; conducting a review of a program or agency; developing a research agenda; evaluating the strengths and weaknesses of a governmental initiative; drawing conclusions and recommendations about a scientific, technical, or policy issue of importance; conducting a public event to inform or highlight a subject; or convening a meeting of disparate experts to address or illuminate a possibly contentious topic.

This report, however, is an intentional departure from these more typical efforts in its approach and intent. As noted above, the Committee on Extraordinary Engineering Impacts on Society was tasked with offering advice to the NSF on how the agency might enhance the public's understanding of the myriad impacts of engineering innovations on their everyday lives and spark interest in the field among the next generation of innovators. The particular focus is the societal impact of NSF's contributions to engineering research—and, by extension, the education of researchers—and the stories of the people who brought those contributions about. Unlike most of the other works cited, the report was directed to be designed for a wide readership.

There are a number of more archetypal National Academies reports, along with selected work done in a similar vein by others, that address topics under the committee's consideration. It was neither the committee's task nor their goal to reproduce these reports. Instead, these publications informed the committee's thinking. They also serve as independent references providing more in-depth explorations of issues dealt with in this volume.

These include reports that:

- evaluate the impact of federally funded research support and, in particular, the technical, economic, and social impact of federally funded information technology research;
- examine the conduct and results of National Science Foundation program initiatives;

- offer advice on communicating information on science, technology, and the value of supporting research to the general public; and
- raise general public awareness of engineering and the role of engineers.

Table 1-1 lists representative publications. Some of the ones that had the most influence on the committee are summarized in subsequent chapters.

TABLE 1-1 Selected National Academies and Other Publications Addressing Topics Under Consideration by the Committee (non-National Academies publications <u>underlined</u>)

Title	Reference
Impacts of Federal Support of Information Technology Research	
Evolving the High Performance Computing and Communications Initiative to Support the Nation's Information Infrastructure	NRC (1995)
Funding a Revolution: Government Support for Computing Research	NRC (2002)
Assessing the Impacts of Changes in the Information Technology R&D Ecosystem: Retaining Leadership in an Increasingly Global Environment	NRC (2009)
Continuing Innovation in Information Technology	NRC (2012b)
Continuing Innovation in Information Technology: Workshop Report	NASEM (2016)
Information Technology Innovation Resurgence, Confluence, and Continuing Impact	NASEM (2020b)
Impacts of Federal Support of Research – General	
The Government Role in Civilian Technology: Building a New Alliance	NAS/NAE/IOM (1992)
Mastering a New Role: Shaping Technology Policy for National Economic Performance	NAE (1993)
Allocating Federal Funds for Science and Technology	IOM/NAS/NAE/NRC (1995)
Rising Above the Gathering Storm: Energizing and Employing America for a Brighter Economic Future.	NAS/NAE/IOM (2007)
Measuring the Impacts of Federal Investments in Research: A Workshop Summary	NRC (2011b)
Capturing Change in Science, Technology, and Innovation: Improving Indicators to Inform Policy	NRC (2014a)
Furthering America's Research Enterprise	NRC (2014b)
NSF Engineering-Related Program Evaluations	
New Directions for Engineering in the National Science Foundation: A Report to the NSF from the National Academy of Engineering	NAE (1985)
The Engineering Research Centers: Leaders in Change	NRC (1987)
<u>*Enabling American Innovation. Engineering and the National Science Foundation*</u>	<u>Belanger (1998)</u>
An Assessment of the SBIR Program at the National Science Foundation	NRC (2008)
SBIR at the National Science Foundation	NASEM (2015)

Title	Reference
A New Vision for Center-Based Engineering Research	NASEM (2017a)
Review of the SBIR and STTR Programs at the National Science Foundation	NASEM (2023c)
Communicating Information on Science, Technology, and the Value of Supporting Research	
Communicating National Science Foundation Science and Engineering Information to Data Users: Letter Report	NRC (2011a)
Communicating Science and Engineering Data in the Information Age	NRC (2012a)
The Science of Science Communications II: Summary of a Colloquium	NAS (2014)
Communicating Science Effectively: A Research Agenda	NASEM (2017b)
The Science of Science Communications III: Inspiring Novel Collaborations and Building Capacity: Proceedings of a Colloquium	NASEM (2018)
Raising Public Awareness of Engineering	
Raising Public Awareness of Engineering	NAE (2002)
Making the Case for Engineering: Study and Recommendations	NSF (2005)
Changing the Conversation. Messages for Improving Public Understanding of Engineering	NAE (2008)
Messaging for Engineering: From Research to Action	NAE (2013)
Agents of Change: NSF's Engineering Research Centers	Preston and Lewis (2020)

Separately, the National Academies have highlighted "engineering achievements that have transformed lives" as part of the "Greatest Engineering Achievements of the 20th Century" initiative (NAE, 2000), later turned into a book (Constable et al., 2003), and NSF itself has produced multiple lists of exceptional inventions, innovations and discoveries that have resulted from their funding efforts (NSF, 2000a,b, 2010, 2022).

An extensive literature on issues related to the societal impact of NSF funding of engineering research and education[2] thus exists, which this report supplements from the unique perspective of the people and stories that underlie the achievements that it has brought forth.

HOW THE COMMITTEE DEFINES "ENGINEERING" AND "IMPACT"

Terms like "engineering" and "impact" have been continuously modified, updated, and debated by professionals and scholars alike over time (see for example, Anderson, 2019; Downey, 2015; Lopez-Cruz, 2022; NSPE, 2018; Weinert, 1986). In conducting this study, the committee applied common definitions of these terms to help conceptualize what engineering is, the distinctions between engineering research and the engineering process, and the scope of societal impacts relevant to its statement of task.

"**Engineering** is the act of creating artifacts, processes, or systems that advance technology and address human needs using principles of the sciences, mathematics, computing, and

---

[2] The National Academies have also produced a number of reports on engineering education in general (NAE, 2005; NAE, 2012; NAE, 2018; NASEM, 2023a) and specifically for K-12 students (NAE and NRC, 2009; NRC, 2010; NASEM, 2020a); topics outside the scope of this report.

operations" (Anderson, 2019). Engineering encompasses not only the design of systems, structures, and devices but also their construction, implementation, deployment, and function.

Engineering research is a systematic investigation conducted to advance knowledge and develop new technologies within the field of engineering. Applied research, an important component of engineering research, aims to devise broadly applicable approaches for practical problems (as distinct from *development*, which focuses on a single product). By building on past designs and learning from failures, engineering research generates knowledge and technologies to continuously improve devices and engineering practice in general, and help the field of engineering adapt to new challenges.

The **engineering process** is a structured approach to designing, developing, and delivering products and services that address specific needs or problems. It involves a series of steps that usually includes problem definition, design, modeling/simulation, prototyping, testing, and implementation, and applies existing knowledge, often based on engineering research, to deploy market-ready solutions. Engineering processes evolve over time as new materials, tools, techniques, and objectives drive innovation.

The **engineering impacts on society** considered by this committee, as set forth in the Statement of Task (Box 1-1) "include expanded technological and social capabilities, scientific breakthroughs, and improvements in economic opportunity. They could have led to improvements in individual quality of life, national security, population health, manufacturing services, infrastructure resilience, and public policy, among others". They were chosen not through quantitative metrics but based on foundational references and the considerations detailed in the report. The goal was to identify societal impacts that were clearly linked to NSF support in some form, and that had the potential to be relatable to a wide audience.

REPORT ORGANIZATION AND FRAMEWORK

The remainder of this report is divided into four additional chapters and three supporting appendices. Chapter 2 delves into the origins of NSF's support for engineering research. It identifies the various divisions, directorates, and other entities that have provided funding over the years and the mechanisms such as Engineering Research Centers that the agency uses to administer and carry out that work. The considerations that informed the committee's evaluation of engineering impacts on society are addressed in Chapter 3. That chapter presents an overview of the topic before touching on the role of NSF in funding engineering innovation and discussing the agency's broader impacts proposal review criterion, which requires consideration of societal impacts. Chapter 4 focuses on recognizing engineering impacts on society brought about by NSF investments. It describes NSF's own initiatives to highlight these impacts and the committee's information-gathering activities via its symposium, questionnaires, and other inputs. The committee's framework for identifying exemplary impacts is then explained, and descriptions of those impacts are presented along with the conclusions and recommendations that were drawn from that work. Where appropriate, attribution is provided to other government agencies or private sector funders who also played significant roles in helping bring about the cited impacts. The last chapter of the report, Chapter 5, addresses the final element of the statement of task: offering advice on communicating engineering's impacts on society to the public. This chapter talks about the considerations that go into communicating impacts and what research tells us about them and enumerates the many audiences there are for such communication. It sets forth the committee's approach to developing example outreach materials, presents summaries of those materials, and closes with the committee's conclusions and recommendations regarding communications issues.

Appendix A reproduces the agenda for the committee's virtual symposium on extraordinary engineering impacts on society. This symposium, which is summarized in proceedings published in 2023 (NASEM, 2023b), provided valuable information and insights to the committee. Appendix B documents the content and motivations behind the draft example outreach materials presented in Chapter 5, including links to online supporting materials. And Appendix C provides biographic information on the committee members and staff responsible for this report.

REFERENCES

Anderson, J. 2019. President's perspective: What Is engineering? *The Bridge* 49(4). https://nae.edu/221278/Presidents-Perspective-What-Is-Engineering (accessed June 4, 2024).

Belanger, D. O. 1998. *Enabling American innovation: Engineering and the National Science Foundation*. Princeton, NJ: Purdue University Press.

Constable, G., and B. Somerville, eds. 2003. *A century of innovation: Twenty engineering achievements that transformed our lives*. Washington, DC: Joseph Henry Press. https://www.nsf.gov/pubs/policydocs/pappg18_1/pappg_3.jsp (accessed February 19, 2024).

Downey, G. L. 2015. PDS: Engineering as problem definition and solution. In S. H. Christensen, C. Didier, A. Jamison, M. Meganck, C. Mitcham, & B. Newberry (Eds.), *International Perspectives on Engineering Education: Engineering Education and Practice in Context* (Vol. 1, pp. 435–455). Springer, Cham. https://doi.org/10.1007/978-3-319-16169-3_21

IOM/NAS/NAE/NRC (Institute of Medicine, National Academy of Sciences, National Academy of Engineering, and National Research Council). 1995. *Allocating Federal Funds for Science and Technology*. Washington, DC: National Academy Press. https://doi.org/10.17226/5040.

Lopez-Cruz, O. 2022. An essential definition of engineering to support engineering research in the twenty-first century. *International Journal of Philosophy* 10(4):130-137. https://doi.org/10.11648/j.ijp.20221004.12.

Margulies S. 2021. *Engineering impacts*. Presentation before the National Academies Committee on Extraordinary Engineering Impacts on Society. Washington, DC. September 27.

NAE (National Academy of Engineering). 1985. *New directions for engineering in the National Science Foundation: A report to the National Science Foundation from the National Academy of Engineering*. Washington, DC: National Academy Press. https://doi.org/10.17226/18892.

NAE. 1993. *Mastering a new role: Shaping technology policy for national economic performance*. Washington, DC: National Academy Press. https://doi.org/10.17226/2103.

NAE. 2000. *Greatest engineering achievements of the 20th century*. http://www.greatachievements.org/ (accessed February 19, 2024).

NAE. 2002. *Raising public awareness of engineering*. Washington, DC: National Academy Press. https://doi.org/10.17226/10573.

NAE. 2005. *Educating the engineer of 2020: Adapting engineering education to the new century*. Washington, DC: The National Academies Press. https://doi.org/10.17226/11338.

NAE. 2008. *Changing the conversation: Messages for improving public understanding of engineering*. Washington, DC: The National Academies Press. https://doi.org/10.17226/12187.

NAE. 2012. *Infusing real world experiences into engineering education*. Washington, DC: The National Academies Press. https://doi.org/10.17226/18184.

NAE. 2013. *Messaging for engineering: From research to action*. Washington, DC: The National Academies Press. https://doi.org/10.17226/13463.

NAE. 2018. *Understanding the educational and career pathways of engineers*. Washington, DC: The National Academies Press. https://doi.org/10.17226/25284.

NAE/NRC (National Academy of Engineering and National Research Council). 2009. *Engineering in K–12 education: Understanding the status and improving the prospects*. Washington, DC: The National Academies Press. https://doi.org/10.17226/12635.

NAS (National Academy of Sciences). 2014. *The science of science communication II: Summary of a colloquium*. Washington, DC: The National Academies Press. https://doi.org/10.17226/18478.

NAS/NAE/IOM (National Academy of Sciences, National Academy of Engineering, and Institute of Medicine). 1992. *The government role in civilian technology: Building a new alliance*. Washington, DC: National Academy Press. https://doi.org/10.17226/1998.

NAS/NAE/IOM. 2007. *Rising above the gathering storm: Energizing and employing America for a brighter economic future*. Washington, DC: The National Academies Press.

NASEM (National Academies of Sciences, Engineering, and Medicine). 2015. *SBIR at the National Science Foundation*. Washington, DC: The National Academies Press. https://doi.org/10.17226/18944.

NASEM. 2016. *Continuing innovation in information technology: Workshop report*. Washington, DC: The National Academies Press. https://doi.org/10.17226/23393.

NASEM. 2017a. *A new vision for center-based engineering research*. Washington, DC: The National Academies Press. https://doi.org/10.17226/24767.

NASEM. 2017b. *Communicating science effectively: A research agenda*. Washington, DC: The National Academies Press. https://doi.org/10.17226/23674.

NASEM. 2018. *The science of science communication III: Inspiring novel collaborations and building capacity: Proceedings of a colloquium*. Washington, DC: The National Academies Press. https://doi.org/10.17226/24958.

NASEM. 2020a. *Building capacity for teaching engineering in K–12 education*. Washington, DC: The National Academies Press. https://doi.org/10.17226/25612.

NASEM. 2020b. *Information technology innovation: Resurgence, confluence, and continuing impact*. Washington, DC: The National Academies Press. https://doi.org/10.17226/25961.

NASEM. 2023a. *Connecting efforts to support minorities in engineering education: Proceedings of a workshop*. Washington, DC: The National Academies Press. https://doi.org/10.17226/27238.

NASEM (National Academies of Sciences, Engineering, and Medicine). 2023. *Extraordinary engineering impacts on society: Proceedings of a symposium*. Washington, DC: The National Academies Press. https://doi.org/10.17226/26847.

NASEM. 2023c. *Review of the SBIR and STTR programs at the National Science Foundation*. Washington, DC: The National Academies Press. https://doi.org/10.17226/26884.

NRC (National Research Council). 1987. *The Engineering Research Centers: Leaders in change*. Washington, DC: National Academy Press. https://doi.org/10.17226/18889.

NRC. 1995. *Evolving the high-performance computing and communications initiative to support the nation's information infrastructure*. Washington, DC: National Academy Press. https://doi.org/10.17226/4948.

NRC. 1999. *Funding a revolution: Government support for computing research*. Washington, DC: National Academy Press. https://doi.org/10.17226/6323.

NRC. 2008. *An assessment of the SBIR program at the National Science Foundation*. Washington, DC: The National Academies Press. https://doi.org/10.17226/11929.

NRC. 2009. *Assessing the impacts of changes in the information technology R&D ecosystem: Retaining leadership in an increasingly global environment*. Washington, DC: The National Academies Press. https://doi.org/10.17226/12174.

NRC. 2010. *Standards for K–12 engineering education?* Washington, DC: The National Academies Press. https://doi.org/10.17226/12990.

NRC. 2011a. *Communicating National Science Foundation science and engineering information to data users: Letter report*. Washington, DC: The National Academies Press. https://doi.org/10.17226/13120.

NRC. 2011b. *Measuring the impacts of federal investments in research: A workshop summary*. Washington, DC: The National Academies Press. https://doi.org/10.17226/13208.

NRC. 2012a. *Communicating science and engineering data in the Information Age*. Washington, DC: The National Academies Press. https://doi.org/10.17226/13282.

NRC. 2012b. *Continuing innovation in information technology*. Washington, DC: The National Academies Press. https://doi.org/10.17226/13427.

NRC. 2014a. *Capturing change in science, technology, and innovation: Improving indicators to inform policy*. Washington, DC: The National Academies Press. https://doi.org/10.17226/18606.

NRC. 2014b. *Furthering America's research enterprise*. Washington, DC: The National Academies Press. https://doi.org/10.17226/18804.

NSF (National Science Foundation). 2000a. *America's investment in the future*. https://www.nsf.gov/about/history/nsf0050/index.jsp (accessed February 19, 2024).

NSF. 2000b. *Nifty 50*. https://www.nsf.gov/about/history/nifty50/index.jsp (accessed February 19, 2024).

NSF. 2005. *Making the case for engineering: Study and recommendations*. https://www.nsf.gov/attachments/104206/public/Final_Case.doc (accessed February 19, 2024).

NSF. 2010. *NSF sensational 60*. https://www.nsf.gov/about/history/sensational60.pdf (accessed February 19, 2024).

NSF. 2022. *NSF history wall*. https://www.nsf.gov/about/history/historywall/NSF_history-guide_July2022_Update.pdf (accessed February 19, 2024).

NSF. 2023. *NSF proposal and award policies and procedure guide*. Chapter III: NSF proposal processing and review. https://new.nsf.gov/policies/pappg/23-1/ch-3-proposal-processing-review (accessed February 19, 2024).

NSPE (National Society of Professional Engineers). 2018. *Defining the practice of engineering*. Alexandria, VA: NSPE.

Preston, L., and C. Lewis. 2020. *Agents of change: NSF's Engineering Research Centers. A history*. https://erc-history.erc-assoc.org/ (accessed February 19, 2024).

Weinert, D. 1986. Appendix A: The definition of engineering and of engineers in historical context. In *Engineering infrastructure diagramming and modeling*. Washington, DC: The National Academies Press. https://doi.org/10.17226/587.

2
NSF's Support of Engineering-Related Research and Education

The National Science Foundation (NSF) is the country's only agency dedicated to the support of fundamental scientific research and is primarily known for this mission. However, since its founding, the NSF has also supported fundamental and applied engineering research and engineering education. This chapter provides an overview of the continual expansion and development of NSF's engineering support, a history that is intertwined with the evolution of the NSF itself. The text begins with a high-level overview of the growth of NSF's support for engineering research and education. It then describes the divisions and mechanisms within the foundation through which this engineering support has been provided.

The text also offers examples of the many societal benefits realized through engineering research and education, with Chapter 4 examining these benefits in greater detail. That chapter's section on "NSF Centers/Engineering Research Centers" details the trajectory of support for engineering at NSF, demonstrating how the changes that NSF has implemented over time to better support engineering have also contributed towards the agency's dual goals of generating knowledge and realizing broader impacts on society.[3]

ORIGINS OF NSF ENGINEERING RESEARCH SUPPORT

Of the $1.1 million that went to research grants in Fiscal Year (FY) 1952, which was the first year that NSF offered research funding,[4] $311,300 was distributed through the Division of Mathematical, Physical and Engineering Sciences, with engineering projects receiving about $42,000 (NSF, 1952; p. 44). The projects supported that year were a study of three-dimensional photoelastic techniques at Brown University, research into fundamental processes in high-voltage breakdown in vacuum at the Massachusetts Institute of Technology, and an examination of the mechanical behavior and structure of linear high polymers at Pennsylvania State College. That same year, 75 of the 624 graduate fellowships awarded by NSF were in engineering (NSF, 1952; p. 22).

Seventy years later, the Engineering Directorate annually distributes more than three-quarters of a billion dollars for engineering research, education, and infrastructure, with other parts of NSF also providing substantial funding for engineering-related endeavors.[5] The result of this seven decades of support for engineering has been an outpouring of new ideas, innovative

---

[3] *Enabling American Innovation. Engineering and the National Science Foundation* (Belanger, 1998) chronicles the early history of this topic in great detail. The ERC Association (https://erc-assoc.org/) maintains an extensive online repository of materials on the Engineering Research Centers program, including an e-book on its history that was completed in 2020 (https://erc-assoc.org/content/erc-program-history-book).

[4] The NSF was established in 1950 but did not start disseminating research support until 1952.

[5] FY 2022 appropriations for NSF are available at https://www.nsf.gov/about/congress/119/highlights/cu22.jsp.

technologies, creative professionals, and dynamic research and educational organizations that have transformed the American economy and society. NSF's evolution to better support funding of engineering research and education has expanded the U.S. economy, benefited human health and the environment, strengthened the nation's security, and supported the education and professional careers of hundreds of thousands of engineers who have been able to pursue their passions while also making the world a better place.

NSF UNITS THAT HAVE SUPPORTED ENGINEERING ACTIVITIES

Support for engineering grew slowly in the 1950s under the Division of Mathematical, Physical and Engineering Sciences established in NSF's founding legislation. By FY 1956, the research grants budget for the division had risen to about $1.4 million, with engineering representing about $726,000 of this amount (NSF, 1956; p. 45). In the next fiscal year, in part due to heightened advocacy from the engineering community, 103 grantees received almost $4.5 million for engineering research (NSF, 1958; p. 54). Among the areas of research singled out in that year's annual report were dynamic and impact studies on concrete beams, the mechanics of information transmission in human speech, and research on the effects of meteor-burst phenomena on the transmission of very high-frequency radio waves.

From the beginning, grants to engineering made by NSF tended not to be made in traditional categories such as civil, mechanical, electrical, chemical, or aeronautical engineering. As the 1952 annual report from the foundation stated, "the emphasis is rather on research fields common to these disciplines, such as fluid mechanics, strength of materials, corrosion, heat transfer, or thermodynamics, because the basic engineering sciences are concerned primarily with the utilization of scientific principles for the general welfare rather than the design aspects of professional engineering. Moreover, the Foundation's program in the engineering sciences and its research support budget is being used to encourage research to fill gaps in the basic information now available to the engineer" (NSF, 1952; pp. 20–21).

NSF also supported cooperative efforts in engineering very early in its history, which would become a prominent theme of the foundation in future years. For example, a cooperative project between mechanical engineers and architects at Princeton University on the effects of a buildings' features on heat control was recommended for funding in 1954, while a Swarthmore College project on neurophysiology combined the fields of biology and electrical engineering (Belanger, 1998). Furthermore, NSF funding of engineering research benefited engineering education as well, both indirectly through the support of engineering faculty and their research efforts and directly through the employment of engineering graduate students to conduct research.

The launch of the Sputnik satellite by the Soviet Union on October 4, 1957, changed the trajectory of engineering research and education within NSF and in the United States in general. The perception that the Soviets had surpassed the United States in critical areas of science and technology drove major increases in investments for engineering research and education. Funding for engineering research within the Division of Mathematical, Physical and Engineering Sciences went from $1.5 million to $4.2 million in one year, and the National Defense Education Act of 1958 authorized federal aid specifically to engineering education (Belanger, 1998). In addition, NSF's involvement with the International Geophysical Year (1957–1958) and other projects involving oceanographic, atmospheric, and astronomical research included substantial engineering components. For example, the foundation made an award of $1.13 million to the

Associated Universities, Inc., for the construction of the National Radio Astronomy Observatory in Green Bank, West Virginia, in 1958 which followed a previous award of $4 million to the observatory (NSF, 1959; p. 41).

The Division of Engineering

The rise of environmental awareness in the 1960s and the increased visibility of pressing national issues in such areas of urban planning, energy production, and transportation further heightened the visibility of engineering as a way of solving national problems. Reflecting the new prominence of the field, in 1964 NSF created a new Division of Engineering separate from the Division of Mathematical and Physical Sciences with a specific mandate to support engineering research and education. As the director of NSF, Leland Haworth, stated, "Today's developments in science and technology, the demands of sociological and cultural change, and the needs of national defense require that basic research in engineering be expanded" (Belanger, 1998; pp. 64–65).

The Engineering Divisional Committee created to advise the new division, which first met the same month that the National Academy of Engineering was established, began to explore areas of engineering not previously supported by NSF, such as oceanographic engineering. NSF funding for engineering the next year went to such areas as plasma dynamics, metals processing, water behavior and management, X-ray microscopy, and holography (NSF, 1966). NSF's growing role in various big science projects, such as the National Center for Atmospheric Research in Colorado and Project Mohole to study the Earth's mantle, had important engineering components.

Interdisciplinary research continued to be a feature of NSF's support of engineering. Grants supporting cooperative work on medical devices involved both engineers and life scientists, while projects in communication science brought together engineers, neurophysiologists, linguists, and others. A grant to a then-new department at the University of Arizona combined geology, agriculture, civil engineering, and atmospheric physics to conduct research in hydrology (NSF, 1966). The growing field of systems design and analysis, buttressed by the growing power of electronic computers, was inherently multidisciplinary and involved engineering in new areas of science and technology.

Graduate enrollments in engineering were increasing at a faster pace than in the physical, life, or social sciences, and the NSF Graduate Traineeships program established in 1964 supported more than 1,200 engineering graduate students that year, with the program later expanding to mathematics, the physical sciences, the biological sciences, and the social sciences (NSF, 1965; p. 68). Research Initiation Grants encouraged early-career faculty at smaller engineering schools to excel in research and education, a form of funding that likewise soon spread to other divisions within NSF. Specialized engineering equipment, conferences, and travel expenses also received funding.

The Research on National Needs Programs

Intensified concern in the tumultuous decade of the 1960s about national problems with links to science and technology led to further examination of how NSF could support research with greater relevance to national needs. Congressional passage of an amended charter for the National Science Foundation, which was signed into law by President Lyndon B. Johnson on July 18, 1968, provided an opportunity to further broaden NSF's mandate. The new charter explicitly cited the social sciences and computer development as disciplines that the Foundation could fund, gave the National Science Board an expanded role in promoting and reporting on scientific research and education, and increased funding for international research cooperation and data gathering. It also specifically sanctioned "applied scientific research relevant to national problems involving the public interest," including engineering studies carried into "early phases of application" (Belinger, 1998; pp. 78–79).

Despite the first budget decline in its history in FY 1969, NSF instituted a new program the following year called Interdisciplinary Research Relevant to the Problems of our Society (IRRPOS). While focused only in part on engineering problems, the IRRPOS program required that proposals explicitly discuss potential societal impacts, even as the Engineering Division continued to support more fundamental engineering research. According to that year's annual report, the IRRPOS program sought "to support interdisciplinary research needed to provide a fuller understanding of major societal problems and to develop new and improved ways to deal with them" (NSF, 1971; p. 55). IRRPOS generated some opposition both inside and outside the foundation from those who worried that it would detract from NSF's primary mission of supporting fundamental research. But calls from policymakers, including members of Congress, the engineering community, and others for more research directed toward national problems demanded a response, and IRRPOS attracted large numbers of research proposals, particularly in the areas of the environment, urban problems, and energy.

In the early 1970s, the Research Applied to National Needs (RANN) program, which soon absorbed IRRPOS, provided another way to fund research directed toward national needs. Although the program only lasted for a few years, it achieved notable accomplishments in such areas as alternative energy sources to supplement fossil fuels, the degradation of shorelines and wetlands, the detection and effects of trace contaminants in the environment, and excavation technologies to accommodate transportation, power, water, and communications systems. As a specific example, work supported by RANN led to the development of wind deflectors for long-haul trucks, which provided fuel savings of 3 to 10 percent (Belanger, 1998; p 106). Like the IRRPOS program, the RANN program encountered resistance regarding its mission and scope, which led to its eventual dissolution. However, it was largely supported by the engineering community; for example, an ad hoc committee of the Committee on Public Engineering Policy at the National Academy of Engineering published a report in 1973, *Priorities for Research Applicable to National Needs*, that addressed the program's research agenda (NAE, 1973).

The Directorate for Engineering

Following the end of the RANN program in the mid-1970s, engineering had several institutional locations at NSF for the next few years, including a reconstituted Directorate for Mathematical and Physical Sciences and Engineering, the Applied Science and Research Applications program (which replaced RANN in 1978), and a new Directorate for Engineering and Applied Science in 1979. Then in 1981, as part of a larger reorganization of the foundation, the "applied sciences" were dropped from the name of that latter directorate, and NSF established the Directorate for Engineering. The directorate was divided into four divisions—Electrical, Computer and Systems Engineering; Chemical and Process Engineering; Civil and Environmental Engineering; and Mechanical Engineering and Applied Mechanics—along with a Problem Analysis Group. Engineering took its place as one of four directorates at the foundation at the time, with the other three covering the biological, behavioral, and social sciences; the astronomical, atmospheric, earth, and ocean sciences; and the mathematical and physical sciences.

In FY 1981, the engineering directorate funded about 1,500 awards with a total expenditure of about $86 million (NSF, 1981). Among the engineering research topics discussed in that year's annual report were robotics, quantum electronics, fusion energy, large-scale networks, renewable energy resources, soil erosion, a new form of welding, hazard mitigation from tsunamis, and the effects of earthquakes on structures.

The elevation of engineering's role within NSF was in part a response to congressional hearings and proposed legislation[6] that would have established a National Technology Foundation, independent from NSF, that would have been similarly organized and managed (Belanger, 1998; Weinschel, 1980). At the time, a major concern in science and technology policy was the international competitiveness of U.S. industry, particularly given the rapid development of the Japanese economy. The mission of the Engineering Directorate was to strengthen the technology programs of NSF while maintaining close connections between engineering research and education and related activities in the sciences. The creation of the new directorate largely satisfied congressional demands to boost the role of engineering and applied research at NSF, and the broad institutional structure for engineering established at that point remains largely intact today. As of early 2024, the Directorate for Engineering—now one of eight directorates within the Foundation—has five major divisions:[7]

1. Division of Chemical, Bioengineering, Environmental and Transport Systems (CBET)
2. Division of Civil, Mechanical and Manufacturing Innovation (CMMI)
3. Division of Electrical, Communications and Cyber Systems (ECCS)
4. Division of Engineering Education and Centers (EEC)
5. Office of Emerging Frontiers and Multidisciplinary Activities (EFMA)

Support for Engineering Research in Other Parts of NSF

The 7 other NSF directorates also support engineering research and education, either directly or indirectly. Support for and deployment of ambitious new specialized research equipment and facilities, including major facilities, often accompanies advances in engineering research and education. These include astronomical observatories such as the Laser Interferometer Gravitational Wave Observatory, national centers such as the National Center for Atmospheric Research, and cooperative international programs in such areas as ocean drilling and polar research.

Three of these directorates have large engineering-related portfolios.

In 1986, NSF's work on computer and information sciences and engineering, which had previously occurred elsewhere in the foundation, was consolidated in its own directorate as an independent administration unit. The mission of the Directorate for Computer and Information Science and Engineering (CISE), as stated on the foundation's website, is "to enable the U.S. to uphold its leadership in computing, communications, and information science and engineering; promote understanding of the principles and uses of advanced computing, communications, and information systems in service to society; support advanced cyberinfrastructure that enables and

---

[6] H.R.6910 — National Technology Foundation Act of 1980; 96th Congress.
[7] https://www.nsf.gov/dir/index.jsp?org=ENG.

accelerates discovery and innovation across all science and engineering disciplines; and contribute to universal, transparent, and affordable participation in an information-based society" (NSF, 2023b). Funding to support this mission goes both to engineering professionals engaged in research and to the training of engineering students.

The Directorate for Science and Engineering Education was established the same year as the Directorate for Engineering, though the Education Directorate was restructured the following year, and its activities were absorbed into the lower-level Office of Scientific and Engineering Personnel and Education. Support for engineering education at NSF, whether distributed through the Engineering Directorate or by another part of NSF, has historically gone to all educational levels and educational settings, including informal education settings such as science museums, educational television, and community events. A particular focus has been broadening the participation of groups that have been underrepresented in engineering by enhancing the quality of engineering education and by demonstrating the relevance of engineering to students' concerns and passions. In 2022, the Directorate for Education and Human Resources, the eventual successor to the Directorate for Science and Engineering Education, was renamed the Directorate for STEM[8] Education. A point of emphasis for the directorate, and particularly its Division of Equity for Excellence in STEM, is to provide support for the "missing millions" of Americans from every background and in every state who have the potential to participate in STEM education and careers.

The Directorate for Technology, Innovation and Partnerships (TIP), created in 2022, seeks to accelerate the translation of research results to the marketplace and society while cultivating new educational pathways that lead to a diverse and skilled future technical workforce. The directorate plans to work collaboratively with all of NSF's other directorates— and with partners in government, industry, philanthropy, civil society, and communities of practice—to take advantage of expertise and resources and energize use-inspired research and innovation. Its priorities are to boost competitiveness, grow the U.S. economy, revitalize communities, and foster a diverse STEM workforce with high-wage jobs. Programs previously housed in other parts of the NSF have been moved to the new directorate, including the Innovation Corps, Partnerships for Innovation, America's Seed Fund, and Pathways to Enable Open-Source Ecosystems (NSF, 2023a). The CHIPS and Science Act, which was signed into law in 2022 (Public Law No. 117-167), authorized[9] $20 billion over 5 years for the TIP directorate. The bill also authorizes the NSF director to identify up to 10 key technology focus areas for the directorate, to be reviewed annually and updated as necessary.

The 4 other NSF directorates also fund some engineering-related activities. Among the Directorate for Biological Sciences' larger current grants is one titled "Integrating engineering theory and biological measures to model stress resilience, damage, and fitness-related consequences";[10] the Directorate for Geosciences is supporting the design and construction of a research submersible;[11] the Directorate for Mathematical and Physical Sciences administers the Materials Research Science and Engineering Centers program,[12] a multi-university effort that facilitates fundamental and applied research and promotes education in that discipline; and the

---

[8] STEM = science, technology, engineering, and math.
[9] An *authorization* establishes the ability for the government to take an action. It does not provide funding (an *appropriation*) for that action.
[10] Award Number: 2015802.
[11] Award Number: 0433409.
[12] https://new.nsf.gov/funding/opportunities/materials-research-science-engineering-centers.

Directorate for Social, Behavioral and Economic Sciences is co-funding a study that is developing technologies that would help people with hearing impairments better manage complex sound environments like a crowded party.[13]

MECHANISMS FOR PROVIDING
ENGINEERING-RELATED SUPPORT AT NSF

NSF has used a wide variety of mechanisms to support engineering research and education, which fall into several broad categories. In recent years, roughly three-quarters of the agency's obligations for research and education programs have been distributed in the form of grants. These grants can be funded as standard awards, where funding for the full duration of the project is provided in a single fiscal year, or as continuing awards, where funding for multiyear projects is provided incrementally. Research grants generally go to the institution for use by the principal investigator for the project proposed, which may include salaries for researchers (including students), the purchase of equipment, and indirect costs retained by the institution.

The remaining one-quarter of NSF's obligations for research and education is distributed in the form of cooperative agreements and contracts. This category of funding is typically used for projects that require substantial agency involvement, as in the case of research centers or multiuse facilities. Contracts are also used to acquire services, studies, or other products that are required for use by NSF or other parts of government.

NSF's FY 2023 financial report estimates that the agency was directly supporting approximately 353,000 researchers, postdoctoral fellows, trainees, teachers, and students (NSF. 2023d). In addition, NSF programs indirectly affect millions of people annually through such activities as educational programs for K–12 students and their teachers, exhibits and programs at informal science institutions such as museums, and other forms of outreach such as television shows, videos, and magazines.

Engineering Research Centers and Other Centers of Excellence

A particular form of funding introduced by the Directorate for Engineering in 1984 and widely adopted elsewhere in government supports the Engineering Research Centers (ERCs), which were designed to conduct multidisciplinary, systems-oriented engineering research on problems critical to industry. ERCs engage in large-scale and long-term programs that span the gamut from transformative basic research to technology development. Their basic goals are to promote cross-disciplinary research, translate research discoveries to innovative products, strengthen the competitiveness of the United States, and firmly link research and education and prepare the next generation of leaders. They are designed to operate simultaneously on three different planes: a systems plane, an enabling technologies plane, and a fundamental knowledge plane. A successful ERC generally receives 10 years of NSF support, after which NSF expects the centers to continue to be operational and impactful. NSF funding for the ERCs has been a maximum of $4 million per year per center, with this limit recently being raised to $6 million per year (NASEM, 2023; p. 44).

Over the program's history to date, there have been four generations of ERCs. Generation 1 (1985–1990) aimed for interdisciplinary, transformational research at a single host university

---

[13] Award Number: 2319321.

with industry engagement. Generation 2 (1994–2006) required the lead university to engage with multiple partner universities, to develop strategic plans showing the pathway from fundamental research to enabling technologies and systems integration, to increase diversity at all levels and include a minority-serving institution, and to expand the educational mission and establish outreach programs to pre-college educational institutions. Generation 3 (2008–2017) sought to transform engineering systems, develop a globally competitive and diverse engineering workforce, and provide cross-cultural, global research, and educational experiences through partnerships with foreign universities and other means. Generation 4 (2020–present), drawing from guidance issued by the National Academies of Sciences, Engineering, and Medicine, emphasizes convergent research and innovation through inclusive partnerships and workforce development (NASEM, 2017a). ERCs are meant to engage a wide range of stakeholders involved in innovation, including students, faculty, staff, leadership, industry, and end users.

As of early 2024, NSF had supported 79 ERCs. A 2022 analysis reported that, "For an NSF investment of less than $2 billion in these centers over 35+ years, the return to the Nation has been estimated at well over $75 billion in new products and processes" (NSF, 2022; p. 3). Over the program's history, ERCs have generated more than 25,000 peer-reviewed journal publications and books, 800 patents, 1,300 licenses, 2,500 invention disclosures, 240 spinoff companies, and more than 14,000 bachelor's, master's, and doctoral degrees earned by ERC students (NASEM, 2023; NSF, 2023c).

The four new centers established in FY 2020 as the first cohort of the fourth-generation ERCs illustrate the program's breadth. They are:
1. Advanced Technologies for Preservation of Biological Systems—the University of Minnesota (lead); Massachusetts General Hospital; the University of California, Berkeley; and the University of California, Riverside
2. Advancing Sustainability through Powered Infrastructure for Roadway Electrification—Utah State University (lead), Purdue University, the University of Colorado, and the University of Texas at El Paso
3. The Internet of Things for Precision Agriculture—University of Pennsylvania (lead); Purdue University; the University of California, Merced; and the University of Florida
4. Quantum Networks—University of Arizona (lead), Harvard University, the Massachusetts Institute of Technology, and Yale University.

In 2022, four additional fourth generation centers were established:
1. Advancing Sustainable and Distributed Fertilizer—Texas Tech University (lead), Case Western Reserve University, Florida A&M University, Georgia Tech, and the Massachusetts Institute of Technology
2. Hybrid Autonomous Manufacturing Moving from Evolution to Revolution—The Ohio State University (lead), Case Western Reserve University, North Carolina Agricultural and Technical State University, Northwestern University, and the University of Tennessee, Knoxville
3. Precision Microbiome Engineering—Duke University (lead), North Carolina Agricultural and Technical State University, North Carolina State University, the University of North Carolina at Chapel Hill, and the University of North Carolina at Charlotte
4. Smart Streetscapes—Columbia University (lead), Florida Atlantic University, Lehman College, Rutgers University, and the University of Central Florida.

The ERCs have acted as a model for other NSF center programs, including the Science and Technology Centers, the Industry-University Cooperative Research Centers, the Earthquake ERCs, and the Nanoscale Science and Engineering Centers. A list of some prominent NSF Centers programs is contained in Table 2-1, along with the year in which they were initiated.

TABLE 2-1 NSF Centers Programs and Their Year of Initiation

Program	Initiation Year
Engineering Research Centers	1985
Science and Technology Centers	1987
Materials Science and Engineering Centers	1994
Centers for Analysis and Synthesis	1995
Centers for Chemical Innovation	1998
Nanoscale Science and Engineering Centers	2001
Artificial Intelligence Research Institutes	2020
Quantum Leap Challenge Institutes	2020
Spectrum Innovation Initiative Centers	2021
NSF Regional Innovation Engines	2023

SOURCE: https://nsf-gov-resources.nsf.gov/2023-03/62_fy2024.pdf.

Other Funding Programs for Engineering-Related Research and Education

The variety of programs through which funding for research and education has been distributed at NSF is too large to inventory in detail in this report. To give just one example: the Research Experience for Undergraduates program is an NSF-wide initiative that provides substantive, real-world engineering and other research experiences. Students generally apply for the program through a competitive process, seeking to spend the summer conducting work in their chosen field, whether in a laboratory or at a field site, either domestically or internationally. The experience typically runs 8–10 weeks and participants are granted stipends and may be provided with housing and travel assistance to facilitate their involvement. (NASEM, 2017; NSF, 2024)

A sense of other past and present programs most closely related to engineering endeavors can be derived from those mentioned by presenters at the committee's 2022 symposium.[14] These are briefly described in Table 2-2; a more complete list may be found on the NSF website.[15]

TABLE 2-2 Examples of NSF Funding Programs Related to Engineering

Program	Description
Broadening Participation in Engineering	Grants to institutions seek to strengthen the future U.S. engineering workforce by enabling and encouraging the participation of all citizens in the engineering enterprise.
Convergence Accelerator	Teams work collaboratively to address societal challenges through convergent research and innovation.

---

[14] https://www.nationalacademies.org/event/08-18-2022/symposium-on-extraordinary-engineering-impacts-on-society.
[15] https://new.nsf.gov/funding.

Program	Description
Experiential Learning for Veterans in Assistive Technology and Engineering	Wounded, injured, and ill veterans receive support to transition into university science, technology, engineering, and mathematics programs, with a special emphasis on assistive technology and engineering.
Faculty Early Career Development	Early-career faculty serve as academic role models in research and education and lead advances in the mission of their departments or organizations.
Grant Opportunities for Academic Liaison with Industry	Academic and industrial researchers work together to generate and share knowledge through research collaborations and fellowships.
Industry–University Cooperative Research Centers	University researchers connect with industry partners through proven frameworks to create and sustain centers that bridge the gap between industry and universities.
Innovation Corps (I-Corps)	Faculty and students take the initial steps in commercializing their discoveries with the help of industry mentors.
Innovation Corps for Learning	An extension of the I-Corps program seeks to advance widespread adoption of promising educational practices to promote STEM learning.
Innovation Technology Experiences for Students and Teachers	Projects that engage students in experiences that increase interest in STEM, conducted under a grant to the American Indian Science and Engineering Society.
Integrative Graduate Education and Research Traineeship	Innovative new transdisciplinary models for graduate education and training prepare a world-class, broadly inclusive, and globally engaged science and engineering workforce.
Major Research Instrumentation	Awards support the acquisition or development of multiuser research scientific and engineering instrumentation.
Partnerships for Innovation	Regional collaborations among academia, industry, and other public and private entities help translate research findings into innovations.
Presidential Early Career Award for Scientists and Engineering	Awards recognize and honor outstanding scientists and engineers at the outset of their independent research careers.
Presidential Young Investigator	Generous base grants renewable for up to 5 years are designed to attract young people to do research in universities and industry.
Research Experience for Undergraduates	Grants to provide summer research programs for undergraduate students that may or may not be from the host institution.
Small Business Innovation Research	Small businesses undertake translational research to develop prototypes and scale up production.
Small Business Technology Transfer	Small businesses collaborate with academic partners on technological innovation.
Small Grant Exploratory Research	Grants for small-scale, exploratory research support high-risk research in science, engineering, and education.

SOURCES: https://new.nsf.gov/funding; NASEM (2023).

As with the ERCs, many of these programs, as well as other programs pioneered by NSF, have acted as models for the establishment of similar programs in other federal agencies. In this way, NSF's programs for research and education have had a widespread influence on academia, government, industry, and nonprofit organizations.

REFERENCES

Belanger, D. O. 1998. *Enabling American innovation: Engineering and the National Science Foundation*. West Lafayette, IN: Purdue University Press.

NAE (National Academy of Engineering). 1973 *Priorities for research applicable to national needs*. Washington, DC: National Academy Press. https://doi.org/10.17226/20383.

NASEM (National Academies of Sciences, Engineering, and Medicine). 2017a. *A new vision for center-based research*. Washington, DC: The National Academies Press. https://doi.org/10.17226/24767.

NASEM. 2017b. *Undergraduate research experiences for STEM students: Successes, challenges, and opportunities*. Washington, DC: The National Academies Press. https://doi.org/10.17226/24622.

NASEM. 2023. *Extraordinary engineering impacts on society: Proceedings of a symposium*. Washington, DC: The National Academies Press. https://doi.org/10.17226/26847.

NSF (National Science Foundation). 1952. *The second annual report of the National Science Foundation, fiscal year 1952*. Washington, DC: U.S. Government Printing Office.

NSF. 1956. *The sixth annual report of the National Science Foundation, fiscal year 1956*. Washington, DC: U.S. Government Printing Office.

NSF. 1958. *The seventh annual report of the National Science Foundation, fiscal year 1957*. Washington, DC: U.S. Government Printing Office.

NSF. 1959. *The eighth annual report of the National Science Foundation, fiscal year 1958*. Washington, DC: U.S. Government Printing Office.

NSF. 1965. *National Science Foundation, fourteenth annual report for the fiscal year ended June 30, 1964*. Washington, DC: U.S. Government Printing Office.

NSF. 1966. *National Science Foundation, fifteenth annual report for the fiscal year ended June 30, 1965*. Washington, DC: U.S. Government Printing Office.

NSF. 1971. *National Science Foundation, twentieth annual report for the fiscal year ended June 30, 1970*. Washington, DC: U.S. Government Printing Office.

NSF. 1981. *National Science Foundation, thirty-first annual report for fiscal year 1981*. Washington, DC: U.S. Government Printing Office.

NSF. 2022. *FY 2020 Engineering Research Centers Program report*. https://www.nsf.gov/pubs/2022/nsf22104/nsf22104.pdf (accessed February 20, 2024).

NSF. 2023a. *About TIP*. https://new.nsf.gov/tip/about-tip (accessed February 20, 2024).

NSF. 2023b. *Directorate for Computer and Information Science and Engineering (CISE)*. https://nsfpolicyoutreach.com/policy-topics-/directorate-computer-information-science-engineering-cise (accessed February 20, 2024).

NSF. 2023c. *Engineering Research Centers*. https://www.nsf.gov/eng/eec/erc.jsp (accessed February 20, 2024).

NSF. 2023d. *FY 2023 agency financial report*. https://www.nsf.gov/pubs/2024/nsf24002/pdf/nsf24002.pdf (accessed February 20, 2024).

NSF. 2024. *Research Experiences for Undergraduates (REU)*.
   https://new.nsf.gov/funding/opportunities/research-experiences-undergraduates-reu (accessed
   June 1, 2024).
Weinschel, B. O. 1980. Politics of technology: Proposal: A National Engineering Foundation:
   Would it not help to halt the slide in U.S. productivity and its loss in share of world markets?
   IEEE Spectrum 17(2):58–60.

3
Considerations in Identifying Engineering Impacts on Society

This chapter outlines the many ways in which engineering research and education impact society, the role that the National Science Foundation (NSF) has played in bringing these impacts about, and the means that it employs to evaluate societal impacts. It begins with an overview of engineering's influences on our everyday lives and then briefly addresses the role of research funding, focusing on NSF. The chapter concludes with a description of the agency's "broader impacts" criterion, a required component of funding applications that addresses a proposed undertaking's "potential to benefit society and contribute to the achievement of specific, desired societal outcomes" (NSF, 2022; p. 3). These considerations informed the development of the committee's approach to identifying engineering impacts made possible by NSF investments, which is addressed in the next chapter.

OVERVIEW – IMPACTS OF ENGINEERING ON SOCIETY

The engineering profession and the engineers it is comprised of have radically influenced our world for the better. Over the past century, engineering innovations such as spacecraft, centralized water treatment, and modern conveniences such as air conditioning have brought ideas to reality and provided for societal needs. For example—as highlighted in the National Academy of Engineering (NAE) publication *A Century of Innovation: Twenty Engineering Achievements that Transformed our Lives* (Constable and Somerville, 2003)—electrification is a foundational engineering improvement. Not only did it change the course of economic development of the United States, but many of the other achievements recognized in the book and cited in this report require electricity and a reliable electrical power grid as prerequisites. Table 3-1 presents a complete list of the innovations cited in the publication.

TABLE 3-1 "The Greatest Engineering Achievements of the 20th Century" Identified by the National Academy of Engineering and a Consortium of Professional Engineering Societies

1. Electrification	11. Highways
2. Automobile	12. Spacecraft
3. Airplane	13. Internet
4. Water supply and distribution	14. Imaging
5. Electronics	15. Household appliances
6. Radio and television	16. Health technologies
7. Agricultural mechanization	17. Petroleum and petrochemical technologies
8. Computers	18. Laser and fiber optics
9. Telephone	19. Nuclear technologies
10. Air conditioning and refrigeration	20. High-performance materials

SOURCE: Constable and Somerville (2003).

Just as engineering solutions advanced society in the past, they will also play an important role in meeting the needs of a future under the pressure of population growth and emerging threats to the public. In 2008 the NAE Committee on Engineering's Grand Challenges identified 14 grand challenges and opportunities for engineering during the world's next few generations which fell into four "broad realms of human concern—sustainability, health, vulnerability, and joy of living" (NAE, 2017; p. 1).[16] The challenges, listed in Table 3-2, have since been taken up by the Grand Challenge Scholars Program, an effort being pursued by engineering schools throughout the United States that is designed to prepare students to be the generation that solves the grand challenges facing society in this century.[17]

TABLE 3-2 The "Grand Challenges for Engineering" Identified in 2008

1. Make solar energy economical	8. Engineer better medicines
2. Provide energy from fusion	9. Reverse-engineer the brain
3. Develop carbon sequestration methods	10. Prevent nuclear terror
4. Manage the nitrogen cycle	11. Secure cyberspace
5. Provide access to clean water	12. Enhance virtual reality
6. Restore and improve urban infrastructure	13. Advance personalized learning
7. Advance health informatics	14. Engineer the tools of scientific discovery

SOURCE: NAE (2008).

New challenges such as those highlighted above require new ways of thinking, and the engineer of the future will need to be trained to meet interdisciplinary challenges that promote the sustainable development of our world (Tabas et al., 2019). They will need to respond to emerging needs of people across all cultures of our globalizing world, which will require insights and knowledge currently held in the social and political sciences. Preparing this type of holistic engineer will require schools to become more appealing to those who have not traditionally entered the engineering profession, including changing the messaging from engineering being not "for everyone" to engineering being a creative avenue that students of all kinds can pursue to improve the world (NAE, 2008).

---

[16] Note that the *NAE Grand Challenges for Engineering* report was originally published in 2008 and updated in 2017.
[17] The NAE Grand Challenges for Engineering website is https://www.engineeringchallenges.org/.

ENGINEERING INNOVATION AND FEDERAL FUNDING

Research underpins the important engineering breakthroughs mentioned in *A Century of Innovation* and NAE's Grand Challenges, and research requires sources of support. While commercial interests fund and conduct most research and development work in the United States, the federal government still plays an important role. In 2019, federal sources accounted for 21 percent of basic and applied research funding (NSF, 2022). The Department of Defense and Department of Health and Human Services (primarily, via the National Institutes of Health) were the largest funders, followed by the Department of Energy, National Science Foundation, NASA, and the Department of Agriculture.

These agencies' research support programs do not operate in isolation, however. Indeed, complementary federal agency support has been instrumental in driving engineering innovation in the United States. This complementary support may take different forms. In some circumstances, differences in mission drive when a particular agency is likely to be the primary support mechanism. The "ARPA's"—the Defense Advanced Research Project Agency, Advanced Research Project Agency–Energy, and Advanced Research Project Agency for Health—are tasked with funding cutting-edge, high-risk research efforts that may, if they show promise, be fostered through their next stage of development by other agencies like NSF. In other cases, particular agencies maintain infrastructure like laboratories and testing facilities that other entities can utilize, saving costs and allowing better utilization of specialized operations and support staff. And research in areas that overlap agency responsibilities may be supported by multiple agencies. Studies that examine health and the built environment, for example, might be cofunded by such diverse federal entities as the U.S. Environmental Protection Agency, National Institutes of Health, Department of Energy, National Institute of Standards and Technology and the Department of Housing and Urban Development. Overall, the complementary support from federal agencies has been a cornerstone of engineering innovation in the U.S., enabling groundbreaking research, fostering cross-sector collaborations, and providing essential resources that promote continuous advancement in engineering.

Three examples of federal funding are cited here to illustrate both the catalyzing role of the NSF and the range of activities it has supported.[18] First, a series of National Academies of Sciences, Engineering, and Medicine (National Academies) reports starting in the mid-1990s illustrated the complex nature of information technology (IT) research and the interdependencies among subfields of computing and communications research (most recently,[19] NASEM, 2020a). This work showed that the IT sector is not self-sufficient and that research support, sometimes taking place over decades, has been crucial to the sector's commercial success. Key projects with agency funding include the creation of NSFNET, a predecessor to the Internet that promoted advanced research and education networking across the United States.

NSF funding has also had an impact in the field of nanotechnology. Generally speaking, the rapid growth of innovation in this field was found to be directly correlated to funding (Huang et al., 2005). Huang and colleagues (2006) found that research funded by NSF and patents authored by NSF-funded researchers demonstrate notably greater influence, as indicated by patent citation metrics over the 2001–2004 timeframe, compared with other groups used for

---

[18] Chapter 2 provides a history of NSF's support of engineering-related research and education and information on the mechanisms that the agency uses to provide that support.

[19] A complete list of the reports in this series is contained in Table 1-1.

comparison. The impact of patents authored by the NSF increasingly expanded over the lifespan of the patent, underscoring the enduring significance of fundamental research in the long term.

As a final example, additive manufacturing—an emerging technology that has revolutionized manufacturing—has greatly benefited from NSF support (IDA Science and Technology Policy Institute, 2013). NSF funded predecessor technologies and supported early advanced manufacturing patents to become proof-of-concepts and prototypes in three major commercial technology areas—binder jetting, powder bed fusion, and sheet lamination (Peña et al., 2014). As advanced manufacturing technologies have developed, NSF has supported research on new processes, novel applications for existing processes, and initiatives related to benchmarking and road mapping.

IMPACTS OF NSF PROGRAMS AND ERCS

As noted in this report, NSF's Engineering Research Centers (ERCs) have had enormous impact, creating strong partnerships with the private sector and universities which foster multifaceted interdisciplinary research and generating thousands of engineering graduates and millions of dollars of economic benefits (NSF, 2015; Roessner et al., 2010). ERCs have helped launch new fields and entirely new systems approaches in areas such as bioengineering and nanosystems (Preston and Lewis, 2020). They have also created disruptive technologies in neurotechnology and biorenewable chemicals by facilitating cross-disciplinary collaboration and industry partnerships. In 2007 the return on NSF investment in ERCs was estimated to be 50:1, and this has since continued to grow (Preston and Lewis, 2020).

NSF-led programs have also provided immense economic benefits to the nation. Having generated more than $9 billion in private investment from 2014 to 2020, NSF's Small Business Innovation Research (SBIR) and Small Business Technology Transfer (STTR) programs have helped small business turn their ideas into marketable products and services (U.S. Congress, 2020). The SBIR program was found to commercialize research at a considerable rate, generate substantial knowledge-based outputs such as patents, and create companies that otherwise would not exist (NASEM, 2015). Led by NSF, the National Nanotechnology initiative (NNI), a highly successful cross-disciplinary and interagency coordination effort, was found to have provided key advancements in research and critical support of responsible development of nanotechnology (NASEM, 2020b). NNI also developed important opportunities for nanotechnology workforce training to strengthen national competitiveness.

The NSF's Designing Materials to Revolutionize and Engineer our Future program produced groundbreaking research that was essential to advancing the White House's Material Genome Initiative (MGI), launched in 2011 (NASEM, 2022). This accelerated the discovery and fast and efficient deployment of advanced materials. PARADIM (Platform for the Accelerated Realization, Analysis, and Discovery of Interface Materials), one of four material innovation platforms associated with MGI, hosted users from 41 universities and national labs and resulted in 140 journal publications from 2016 to 2021 (Nutt, 2021). The platform's most notable discoveries include a new type of topological insulator[20] and a form of galfenol (an alloy of iron and gallium) that is the world's highest-performance magnetostrictive material. These impacts

---

[20] A topological insulator is a substance that has an interior that exhibits the properties of an electrical insulator and a surface that acts as a conductor.

are just a few examples among the multitude of impacts of NSF programs that spur engineering innovations that benefit the economy.

THE NSF BROADER IMPACTS REVIEW CRITERION

Overview

A core component of proposals submitted to NSF is the broader impacts criterion, defined as the potential of the research to benefit society and contribute to the achievement of specific desired societal outcomes (NSF, 2019). Currently NSF is the only federal agency that has such a requirement for its proposals. From 1981 to 1997, NSF proposals were evaluated on (1) researcher performance competence, (2) intrinsic merit of research, (3) utility of research, and (4) effect of research on infrastructure of science and engineering. However, in 1997 a reassessment by the National Science Board distilled the four criteria into two evaluation categories: intellectual merit and broader impacts (NSF, 1997). The change came from a confluence of factors including recommendations from the Committee on Equal Opportunities in Science and Engineering, the passing of the Government Performance and Results Act, and the "NSF in a Changing World" strategic plan (NSF, 1995) which included a long-term goal of promoting knowledge in service of society.

While NSF avoids dictating the exact societal outcomes a project should target, it aims to ensure that public funding supports research with tangible societal benefits beyond fundamental knowledge expansion. The NSF website[21] offers various examples of Broader Impacts Criteria (BIC), which are illustrative but not exhaustive, including:

- inclusion;
- science, technology, engineering, and mathematics (STEM) education;
- public engagement;
- societal well-being;
- STEM workforce development;
- partnerships between academia, industry, and others;
- national security;
- economic competitiveness; and
- infrastructure for research and education.

Prior iterations have categorized broader impacts as: (1) infrastructure for science, (2) broadening participation, (3) training and education, (4) academic collaboration, (5) K–12 outreach, (6) potential society benefits, (7) outreach/broad dissemination, and (8) partnerships with potential users of research results (Roberts, 2009, based on a July 2007 NSF broader impacts guidance memorandum cited in the paper); other impacts have included (1) increased public scientific literacy, (2) increased public engagement with science and technology, (3) broadened participation, (4) development of a diverse STEM workforce, (5) development of a globally competitive STEM workforce, and (6) increased economic competitiveness of the United States (Verdín, 2017; based on NSF's 2016 *Proposal & Award Policies and Procedures*

---

[21] https://new.nsf.gov/funding/learn/broader-impacts.

Guide). NSF's initiatives are widespread in both subject matter and programmatic nature, encompassing broader impacts that range from those inherent to the research itself to broadening participation benefits from STEM outreach programs.

A variety of efforts have been undertaken to evaluate and address NSF's broader impact criterion. These include a study conducted by the National Academy of Public Administration in coordination with NSF that identified the need for a more consistent application of broader impacts as a criterion during proposal reviews (NAPA, 2001). In 2010, Congressional approval of the America Creating Opportunities to Meaningfully Promote Excellence in Technology, Education, and Science (COMPETES) Reauthorization Act[22] mandated broader impacts and ordered higher education institutions to support principal investigators achieving broader impacts in their work.

In 2013 the National Alliance for Broader Impacts (NABI) was formed through funding from an NSF Research Coordination Network grant. Made up of approximately 200 institutions from across the United States, NABI was created to develop institutional capacity in STEM fields to respond to the broader impacts criterion. The organization developed a guiding document in 2015 for reviewers that was the first nation-wide attempt to standardize how review panels assess NSF proposals' broader impact plans. A follow up NABI report in 2018 assembled data from years of its annual summits and two NSF reports on broader impacts implementation and application across directorates. The findings of this report highlighted seven issues common to all relevant stakeholder groups to better inform how NSF's broader impacts criterion evaluation should be carried out. These issues were largely tied to the lack of clarity and consistency of broader impact evaluation as well as a lack of resources to support broader impacts at the individual, institutional, and national level. The document, *The Current State of Broader Impacts: Advancing Science and Benefiting Society,* is listed as one of their key additional resources to stakeholders today (NABI, 2018).

Evaluations of the Effects of Broader Impacts Reviews

Over recent years, scholars have assessed how broader impacts have been distributed among the categories described in the previous section (Kamenetzky, 2013). The most frequent categories included were teaching and training, broad dissemination, and infrastructure enhancement. Verdín (2017) found that broader impact statements were most likely to include (1) increased public scientific literacy, (2) public engagement in science and engineering, and (3) developing a diverse STEM workforce. Watts et al. (2015), on the other hand, found that activities seeking to broaden participation of underrepresented groups in STEM fields were reported less frequently. Cultural differences between scientific fields have been found to play a large role in how these fields view the broader impacts criterion, causing them to propose certain types of impacts in their submissions (Kamenetzky, 2013). Political considerations have also been proposed as influencing the types of broader impacts that researchers mention or omit in their award proposals (Roberts, 2009).

Researchers who submitted proposals that mentioned benefits to society were found to be no more likely to propose dissemination of their results to those who could use them than researchers who only spoke to the broader impacts of their work for science (Roberts, 2009).

---

[22] Public Law 111-358.

This finding suggests that the potential societal benefits identified when describing broader impacts do not necessarily lead to actual societal benefits and that results relevant to the public are often not publicized outside of the scientific community. Actions are thus needed to communicate broader impacts along with dissemination activities to both researchers and the general public to build a better knowledge base regarding societal impacts, especially among underrepresented groups.

The creation of NSF's broader impacts criterion along with agency and university efforts to create institutional capacity around them have led to important progress in linking research projects and societal impacts. However, it was found that the current broader impacts framework could be improved by assessing who receives the benefits created by NSF-funded research (Bozeman, 2020) and by establishing metrics for broader impact assessment (Verdín, 2017). Without such an analysis, NSF-funded research could sustain or even exacerbate inequalities that exist today (Woodson et al., 2021). The Inclusion-Immediacy criterion (IIC) created by Woodson and Boutilier (2022) seeks to fill this gap and shed light on the impacts of funded NSF projects on marginalized communities. IIC evaluates NSF grants based on the people who will benefit from the research and characterizes the grant based on the alignment of the research and the broader impacts. Using this framework has the potential to benefit not only marginalized communities but the entire innovation ecosystem, as inequality limits NSF's mission to advance national health, prosperity, welfare, and security (NSF, n.d.).

REFERENCES

Bozeman, B. 2020. Public value science. *Issues in Science and Technology* 36(4):34–41.

Constable, G., and B. Somerville. 2003. *A century of innovation: Twenty engineering achievements that transformed our lives.* Washington, DC: Joseph Henry Press. https://doi.org/10.17226/10726 (accessed February 22, 2024).

Huang, Z., H. Chen, L. Yan, and M. Roco. 2005. Longitudinal nanotechnology development (1991–2002): National Science Foundation funding and its impact on patents. *Journal of Nanoparticle Research* 7:343–376. https://doi.org/10.1007/s11051-005-5468-3 (accessed February 22, 2024).

Huang, Z., H. Chen, X. Li, and M. C. Roco 2006. Connecting NSF funding to patent innovation in nanotechnology (2001–2004). *Journal of Nanoparticle Research* 8(6):859–879. https://doi.org/10.1007/s11051-006-9147-9 (accessed February 22, 2024).

IDA Science and Technology Policy Institute. 2013. *The role of the National Science Foundation in the origin and evolution of additive manufacturing in the United States p-5091).* Institute for Defense Analyses. https://www.ida.org/-/media/feature/publications/t/th/the-role-of-the-national-science-foundation-in-the-origin-and-evolution-of-additive-manufacturing-in/ida-p-5091.ashx (accessed February 22, 2024).

Kamenetzky, J. R. 2013. Opportunities for impact: Statistical analysis of the National Science Foundation's broader impacts criterion. *Science and Public Policy* 40(1):72–84. https://doi.org/10.1093/scipol/scs059 (accessed February 22, 2024).

NABI (National Alliance for Broader Impacts). 2018. *The current state of Broader Impacts: Advancing science and benefiting society.* https://researchinsociety.org/wp-content/uploads/2021/02/NabiCurrentStateOfBI-011118.pdf (accessed February 22, 2024).

NAE (National Academy of Engineering). 2008. *Changing the conversation: Messages for improving public understanding of engineering.* Washington, DC: The National Academies Press. https://doi.org/10.17226/12187 (accessed February 22, 2024).

NAE. 2017. *NAE grand challenges for engineering.* Washington, DC: The National Academies Press. https://nae.edu/187212/NAE-Grand-Challenges-for-Engineering. (Note that this report was originally published in 2008 and updated in 2017.)

NAPA (National Academy for Public Administration). 2001. A study of the National Science Foundation's criteria for project selection. https://citeseerx.ist.psu.edu/document?repid=rep1&type=pdf&doi=dbeb6cd123afda1f773694 90fd27ffcb3e2df6d9 (accessed February 22, 2024).

NASEM (National Academies of Sciences, Engineering, and Medicine). 2015. *SBIR at the National Science Foundation.* Washington, DC: The National Academies Press. https://doi.org/10.17226/18944 (accessed February 22, 2024).

NASEM. 2020a. *Information technology innovation: Resurgence, confluence, and continuing impact.* Washington, DC: The National Academies Press. https://doi.org/10.17226/25961 (accessed February 22, 2024).

NASEM. 2020b. *A quadrennial review of the National Nanotechnology Initiative: Nanoscience, applications, and commercialization.* Washington, DC: The National Academies Press. https://doi.org/10.17226/25729 (accessed February 22, 2024).

NASEM. 2022. *NSF efforts to achieve the nation's vision for the Materials Genome Initiative: Designing Materials to Revolutionize and Engineer Our Future (DMREF).* Washington, DC: The National Academies Press. https://doi.org/10.17226/26723 (accessed February 22, 2024).

NSF (National Science Foundation). n.d. *About NSF.* https://beta.nsf.gov/about (accessed February 8, 2023).

NSF. 1995. *NSF in a changing world.* https://www.nsf.gov/pubs/1995/nsf9524/contents.htm.

NSF. 1997. *NSF to adopt new merit review criteria.* https://www.nsf.gov/news/news_summ.jsp?cntn_id=102789 (accessed February 22, 2024).

NSF. 2015. *NSF engineering research centers: Creating new knowledge, innovations, and technologies for over 30 years (No. 15–810).* Arlington, VA: National Science Foundation.

NSF. 2019. *Proposal and award policies and procedures guide.* https://www.nsf.gov/pubs/policydocs/pappg19_1/index.jsp (accessed February 22, 2024).

NSF. 2022. *Perspectives on Broader Impacts.* https://nsf-gov-resources.nsf.gov/2022-09/Broader_Impacts_0.pdf (accessed February 22, 2024).

Nutt, D. 2021. $22.5M NSF grant accelerates materials discovery. *Cornell Chronicle,* May 25. https://news.cornell.edu/stories/2021/05/225m-nsf-grant-accelerates-materials-discovery (accessed February 22, 2024).

Peña, V., B. Lal, and M. Micali. 2014. U.S. federal investment in the origin and evolution of additive manufacturing: A case study of the National Science Foundation. *3D Printing and Additive Manufacturing* 1(4):185–193. https://doi.org/10.1089/3dp.2014.0019 (accessed February 22, 2024).

Preston, L., and C. Lewis. 2020. *Agents of change: NSF's engineering research centers—A history.* Lynn Preston Associates LLC. https://erc-history.erc-assoc.org/.

Roberts, M. R. 2009. Realizing societal benefit from academic research: Analysis of the National Science Foundation's broader impacts criterion. *Social Epistemology* 23(3–4):199–219. https://doi.org/10.1080/02691720903364035 (accessed February 22, 2024).

Roessner, J., L. Manrique, and J. Park. 2010. The economic impact of engineering research centers: Preliminary results of a pilot study. *Journal of Technology Transfer* 35:475–493. https://doi.org/10.1007/s10961-010-9163-x (accessed February 22, 2024).

Tabas, B., U. Beagon, and K. Kövesi. 2019. *Report on the future role of engineers in society and the skills and competences engineering will require.* https://arrow.tudublin.ie/engschcivrep/16 (accessed February 22, 2024).

U.S. Congress. 2020. America's seed fund: A Review of SBIR and STTR. Hearing before the Subcommittee on Research and Technology, Committee on Science, Space, and Technology, House of Representatives, One Hundred Sixteenth Congress, second session, February 5, 2020. https://www.congress.gov/event/116th-congress/house-event/LC65772/text (accessed March 20, 2024).

Verdín, D. 2017. Quantifying and assessing trends on the National Science Foundation's broader impact criterion. *2017 ASEE Annual Conference & Exposition Proceedings*: 28778. https://doi.org/10.18260/1-2--28778.

Watts, S. M., M. D. George, and D. J. Levey. 2015. Achieving broader impacts in the National Science Foundation, Division of Environmental Biology. *BioScience* 65(4):397–407. https://doi.org/10.1093/biosci/biv006 (accessed February 22, 2024).

Woodson, T., and S. Boutilier. 2022. Impacts for whom? Assessing inequalities in NSF-funded broader impacts using the inclusion–immediacy criterion. *Science and Public Policy* 49(2):168–178. https://doi.org/10.1093/scipol/scab072 (accessed February 22, 2024).

Woodson, T. S., E. Hoffmann, and S. Boutilier S. 2021. Evaluating the NSF broader impacts with the inclusion–immediacy criterion: A retrospective analysis of nanotechnology grants. *Technovation* 101(C) https://ideas.repec.org//a/eee/techno/v101y2021ics0166497220300821.html (accessed February 22, 2024).

4

Recognizing Engineering Impacts on Society Brought About by NSF Investments

The committee's statement of task directed it to identify up to ten extraordinary engineering impacts made possible by National Science Foundation (NSF) investments; to organize a public symposium on this topic; and to develop clear, compelling narratives for public engagement in these impacts. This chapter addresses these three elements. It describes NSF's own initiatives to highlight the effects that its work has had on science, technology, and society. The committee's efforts to gather information are then presented. These include a symposium—summarized here and detailed in a separate proceedings (NASEM, 2023b)—that not only highlighted the many innovations that resulted from NSF support of engineering research and education but also brought to light the persons responsible for those innovations and the stories of how they came about. Other committee initiatives—questionnaires circulated to the members of the National Academy of Engineering and input received from NSF staff—are then discussed.

The chapter goes on to explain the committee's framework for identifying exemplary impacts and presents descriptions of these impacts. It closes with the conclusions and recommendations that were drawn from this work.

CONSIDERATIONS IN RECOGNIZING THE WAYS IN WHICH ENGINEERING RESEARCH, PRACTICE, AND EDUCATION AFFECT SOCIETY

The study's statement of task identified several different ways that engineering research, practice, and education might have an impact on society. It noted that

> [t]hese impacts might include expanded technological and social capabilities, scientific breakthroughs, and improvements in economic opportunity. They could have led to improvements in individual quality of life, national security, population health, manufacturing services, infrastructure resilience, and public policy, among others.

In formulating its own approach to the question and considering the role of NSF support, the committee was mindful of the fact that significant societal impacts often stem from multiple sources and progress over time—they neither happen instantaneously nor are they the result of a single action or agent. This is especially true when considering the impact of federal funding efforts. Federal agencies have different missions, constituencies, roles, and support practices. Their actions may take the form of funding fundamental, often speculative, research that develops the basic principles that underlie an entire field, nurturing promising but undeveloped technologies until they mature to the point of independent viability, or providing long term support to efforts that generate great social value but not necessarily a high economic return. NSF is unusual among federal agencies in that it operates in all of these spheres, and the committee sought to recognize all of them. It was also careful to acknowledge the other governmental and non-governmental contributors to the impacts it chose to highlight.

PREVIOUS EFFORTS TO IDENTIFY IMPACTS BROUGHT ABOUT BY NSF INVESTMENTS

The National Science Foundation has undertaken three efforts over recent years to identify and document the effects of governmental support for research and education on the economy, human well-being, and societal advancement. These efforts—which coincided with the 50th, 60th, and 70th anniversaries of the founding of the agency—highlighted the substantial return on investment of government support for science and technology and served as resources to inform the committee's considerations of engineering impacts on society.

NSF *Nifty 50*

Developed for the 50th anniversary of agency in 2000, the *Nifty 50* are "NSF-funded inventions, innovations and discoveries that have become commonplace in our lives" (NSF, 2000c). Nearly half of the list, which covers the breadth of NSF's support initiatives, cites inventions, technologies, or programs related to engineering. Table 4-1 displays these.

TABLE 4-1 NSF *Nifty 50* (2000) Achievements Most Directly Related to Engineering

- Bar codes
- Buckyballs
- CAD/CAM
- Computer visualization techniques
- The Darci Card[23]
- Data compression technology
- Doppler radar
- Earthquake mitigation
- "Eye chip" or retina chip
- Fiber optics
- The internet
- MRI: magnetic resonance imaging
- MEMS: microelectromechanical systems
- Nanotechnology
- The Partnerships for Advanced Computational Infrastructure (PACI) program
- Persons with disabilities access to the web
- Reaction injection molding
- Speech recognition technology
- Tissue engineering
- vBNS: very high speed backbone network system
- Volcanic eruption detection
- Web browsers

SOURCE: NSF (2000c).

On the *Nifty 50* website, each of these items is accompanied by a description of how NSF's support contributed to the development. For example, NSF's contributions to bar codes are described as follows:

NSF funding helped play a crucial role—both earlier and more recently—in the development of bar codes. In the early 1990s, research in computer vision conducted at the State University of New York–Stony Brook led to major advances in algorithms for bar code readers. That research led to commercial development of a new product line of barcode readers that has been described

---

[23] The Darci Card is an adaptive tool that empowers individuals with physical disabilities to operate a computer using either an on-screen keyboard or Morse code input methods.

as a revolutionary advance, enabling bar-code readers to operate under messy situations and adverse conditions. Work continues on developing two-dimensional bar codes,[24] which will enable far greater amounts of information to be represented in a very compact form. NSF helped fund bar-code research in the 1970s, which helped to perfect the accuracy of the scanners that read bar codes. Credit must also be given to private industry for its work in development and implementation of bar codes and scanners. (NSF, 2000c)

NSF *Sensational 60*

Ten years later, for its 60th anniversary, NSF produced a list of 60 advances—the so-called *Sensational 60*—that "have had a large impact or influence on every American's life" (NSF, 2010b). The items on this new list that were not among the Nifty 50 included a number of engineering related innovations:

- biofuels and clean energy
- cloud computing
- deep-sea drilling
- (NSF-funded) engineering research and science and technology centers
- functional magnetic resonance imaging
- Google
- "invisibility cloaks"
- RSA[25] and public-key cryptography
- supercomputer facilities

These advances, too, are accompanied by descriptions of NSF's involvement in the research. For functional magnetic resonance imaging (fMRI)), for example, the description is:

> Since the development of fMRI in the early 1990s, NSF has supported numerous fMRI studies that have resulted in a deeper understanding of human cognition across a spectrum of areas, including sensation, perception, motor control, memory, empathy, language, and emotion. Additionally, NSF has funded the development of new statistical methods necessary for the analysis and interpretation of fMRI data. In 1999, NSF helped launch the National Functional Magnetic Resonance Imaging Data Center (fMRIDC) at Dartmouth University. Now funded by the National Institutes of Health and maintained at the University of California, Santa Barbara, the fMRIDC provides a publicly accessible repository of studies and data to further the study of the human brain. fMRI is an excellent example of how NSF funding supports technological innovation and medical advances. NSF supported the underlying NMR, as well as research in other areas directly related to the development of MRI technology, such as electromagnetics, digital systems, computer engineering, biophysics and biochemistry (NSF, 2010b; p. 25).

NSF "History Wall"

More recently, for the NSF's 70th anniversary, NSF commissioned a mural for its new headquarters in Alexandria, Virginia, that provides a visual history of the effects research supported by the foundation has had on society (Figure 4-1) (NSF, 2022f). Created by artist

---

[24] For example, QR (Quick Response) codes, which can be scanned by a smart phone and used to open webpages, call phone numbers, and the like.
[25] The Rivest-Shamir-Adleman (RSA) encryption algorithm is a widely used means to secure data.

Nicolle R. Fuller, the History Wall includes many items on the above lists but also several other items. These are presented in a descriptive format and include:

- Carbon nanotubes have novel properties yielding new applications.
- Geckos inspire the development of polymers and directional adhesion materials.
- NSF-funded search-and-rescue robots improve disaster response.
- NSF supports GPS technology, such as the National Center for Geographic Information and Analysis.
- NSF's SBIR program strengthens the role of small business in federally funded R&D, as it did in cellular technology in the 1990s.
- In electronics and material science, graphene's unique electrical and physical properties promise new breakthroughs.
- Robobees are innovative autonomously flying microrobots that have potential impacts in many applications.
- Quantum phenomena can yield novel technologies in computing and communications.
- 3D printing has impacted manufacturing, design and the arts.
- With support for programs like "The Magic School Bus," NSF supports elementary and informal STEM education.
- Robotics and automation promise to transform transport and more.
- NSF supports potentially transformative technologies like virtual reality.

48

FIGURE 4-1 The History Wall commissioned for NSF's headquarters, compiling images of the effects of NSF-sponsored research on society.
SOURCE: NSF (2022).

THE COMMITTEE'S OUTREACH EFFORTS TO IDENTIFY IMPACTS BROUGHT ABOUT BY NSF INVESTMENTS

The committee undertook two outreach efforts intended to supplement their own knowledge of engineering innovations brought about by NSF investments and the background research conducted by staff. One of these was a symposium carried out in fulfillment of the statement of task. The other was a set of questionnaires circulated to members of the National Academy of Engineering, along with a companion solicitation of input from NSF staff. These efforts are described below.

Symposium on Extraordinary Engineering Impacts on Society

The committee organized and conducted a virtual information-gathering symposium in August 2022. The 2-day event comprised four sessions that touched on major themes raised in the statement of task:

1. NSF and its role in fostering extraordinary engineering impacts on society
2. People who brought about extraordinary engineering impacts on society
3. NSF centers that catalyzed extraordinary engineering impacts on society
4. NSF processes that fostered extraordinary engineering impacts on society

It featured two keynote speakers, 23 session presenters, and an additional three discussants. The event was webcast and attracted logins from nearly 600 unique IP addresses from across the United States and 22 other countries.

The symposium speakers provided many examples of how support of investments in engineering education and research by NSF has led to positive societal and economic impacts. These are listed in Table 4-2.

TABLE 4-2 Engineering Impacts Brought About by NSF Investments Cited by Participants in the Symposium on Extraordinary Engineering Impacts on Society

- 3D printing
- Biomimetic microelectronic systems to restore vision to the blind
- Carbon nanotubes as thermal interface materials
- Computer-aided integrated circuit manufacturing
- Digital research and education networks
- Equipment to assess the performance of engineered devices
- Expansion of the internet
- Fast-switching color shutter
- Gel casting processes using nanoparticles for microdevice production
- Large-scale earthquake shake tables to test structures
- Microbial processes to produce antimalarial drugs
- Nano-reinforced polymer composites
- Novel bioreactors to support cell cultivation
- Open-source software for database-driven commerce
- Performance-based earthquake engineering frameworks
- Polyelectrolyte nanoparticles as drug delivery mechanisms
- Self-powered devices to manage chronic diseases
- Synthetic biology
- Tests to detect graft-versus-host disease
- Tools to probe the hidden areas of unexplored regions
- Tools to understand the electronic structure of carbon compounds
- Trustworthy information systems for cyber infrastructure
- User interfaces for assistive robotic manipulators
- Web browsers

SOURCE: NASEM (2023b).

The symposium had 3 primary purposes. First, it highlighted the rich interconnections between science and engineering and the many ways in which each informs the other. An innovation in one area can have unexpected and far-reaching implications in other areas, and these lines of influence extend in all directions. With semiconductor technology, to take just one example, expanded research often follows technological innovation and is not just a precursor. The linear model—scientific research leading to technological development—upon which NSF was founded has given way to a much richer and more complex picture of how science, technology, and innovation are intertwined.

Second, the symposium emphasized the deep and productive link between research and education. When undergraduate and graduate students, postdoctoral fellows, and early-career researchers learn, graduate, and move to new jobs, they carry with them not just knowledge and expertise but networks of human connections. Gradually, their networks of knowledge, projects, and community broaden and deepen, creating cultural as well as intellectual, social, and human capital. As they advance in their careers, these individuals participate in institutional, disciplinary, and policy-making entities through which they can direct the enterprise in productive directions. In the process, they help to create new companies, new products, and new industries that bolster national security, economic competitiveness, and human health and well-being.

Third, the symposium demonstrated that an underappreciation or misunderstanding of engineering is partially responsible for hindering both technological progress and the participation of underrepresented groups that could contribute to engineering advancements as well as gain benefits from their participation. Greater understanding of the role of engineering and engineers in society could spur increased interest and engagement in the engineering professions and raise the visibility of engineering as a viable career path in diverse communities.

Extraordinary Engineering Impacts on Society. Proceedings of a Symposium (NASEM, 2023b) documents the event in detail. Videos of the presentations and discussions are also posted to the web for viewing, along with copies of the presentation slides.[26]

Questionnaires Circulated to National Academy of Engineering Members and Input Received from NSF Staff

The committee circulated two questionnaires to gather insights into the perceived societal impacts resulting from NSF support of engineering research and education and to inform their own evaluation of impacts. These were sent to members of the National Academy of Engineering. NAE members are elected by their peers in recognition of their achievements in "business and academic management, in technical positions, as university faculty, and as leaders in government and private engineering organizations" (NAE, 2023a). They were invited via email to respond to a short questionnaire, the primary question of which was:

> Do you have knowledge of any significant engineering impacts on society resulting from funding provided by the National Science Foundation?

The questionnaires were circulated in November 2021 and again in February 2023. Table 4-3 presents the combined responses (n=87). Submissions are categorized according to the engineering domains outlined by the NAE's disciplinary engineering sections,[27] with an additional category for Education and Workforce Development. It is important to note that many of the impacts cited span multiple engineering disciplines and the categorizations reflect the committee's judgement about the most representative discipline[s]. NAE sections where no impacts were reported were omitted from the table.

TABLE 4-3 NAE Member Questionnaire Responses—Significant Engineering Impacts on Society

Engineering Discipline	Engineering Impact[s]
Aerospace	• Doppler radar for microburst detection
Bioengineering	• Cloning • CRISPR (×2) • Advanced drug delivery and controlled release systems • Improvements in health care from palliative care to restorative approaches, e.g., precision, noninvasive diagnostics • Hollow fiber membranes • Center for Biofilm Engineering (ERC[28])
Chemical	• Membrane separations for desalinating water, separating air, and increasing separation capacity
Civil & Environmental	• Natural disaster (earthquake) mitigation

---

[26] These materials may be found at https://www.nationalacademies.org/event/08-18-2022/symposium-on-extraordinary-engineering-impacts-on-society.

[27] A complete listing of NAE sections may be found at https://www.nae.edu/166166/Sections.

[28] Engineering Research Center.

Engineering Discipline	Engineering Impact[s]
	• Use of Hurricane Katrina data to mitigate/prevent future major flooding • Improved understanding in dynamic response analysis for embankment dams, e.g., wave propagation (×2) • Collaborations with industry leading to advances in large-scale transportation network dynamics, including: • Reliability and resilience of transportation networks and logistics systems, designing effective evacuation routes, behavioral economics, and the role of electronic marketplaces
Computer Science & Engineering	• NSF Supercomputing Centers program (×2) • Optimal control theory, Kalman filtering, dynamic programming • Mosaic (graphical web browser) • Technologies/algorithms leading to Google, e.g., PageRank (×6) • Akamai (content delivery networks) • OpenFlow (software-defined networking) (×2) • Machine learning (deep learning and natural language processing) • Relational database systems (INGRES) • NSFNET (×2)/CSnet/the internet • Model-checking technology for semiconductor design—The Metal Oxide Semiconductor Implementation Service (MOSIS) • Understanding and generating language from multilingual speakers, including low-resource languages and code-switched speech • Duolingo (language learning app) • Public key cryptography (×2)/ homomorphic encryption • Cybersecurity innovations/differential privacy/secure multi-party computation • Virtual reality
Electric Power/Energy Systems	• Practical AC photovoltaic panels
Electronics, Communication & Information Systems	• Magnetic storage density improvements • DSL/broadband-access • Compressed sensing (signal processing) • Symbolic trajectory evaluation • Circuit and system simulation • CubeSats • Berkeley Sensor & Actuator Center (IUCRC[29]) • Center for Wireless Integrated MicroSensing and Systems (ERC)
Industrial, Manufacturing & Operational Systems	• Reconfigurable manufacturing systems • MEMS (micro-electro-mechanical Systems) fabrication technologies • Center for Plasma-Aided Manufacturing (ERC)
Mechanical	• New paradigm for engineering fluid flows
Education and Workforce Development	• Graduate Research Fellowship Program • SBIR (Small Business Innovation Research) program • Global scientific leadership: "The U.S. would not lead the world in fundamental research in all fields of science without [NSF funding]"

---

[29] Industry-University Cooperative Research Centers.

NOTES: Similar responses are grouped together. Notations—(×2), for example—indicate the number of times that the impact was cited by respondents.

In addition, input was solicited from NSF staff on the same question. Table 4-4 summarizes the responses received (n=6), using the same disciplinary breakouts as above.

TABLE 4-4 NSF Staff Input—Significant Engineering Impacts on Society

Engineering Discipline	Engineering Impact[s]
Bioengineering	• Biomimetic MicroElectronic System Center (ERC) • Center for Neuromorphic Systems Engineering (ERC) • Proof-of-concept development of a cost-effective, room temperature stable means of vaccination • Integrating femtosecond lasers into Lasik • Center for Neurotechnology (ERC)
Civil & Environmental	• Center for Reinventing the Nation's Urban Water Infrastructure (ERC)
Computer Science & Engineering	• Support for research leading to the development of the Viterbi Decoder on a single chip • AI and machine learning (knowledge representation and reasoning, games, expert systems, probabilistic AI, affective computing, robotics, machine learning)
Electric Power/Energy Systems	• Photovoltaic research
Electronics, Communication & Information Systems	• Wide bandgap semiconductors for power electronics and light emitting diodes • Quantum information science and engineering • Perpendicular and heat-assisted magnetic recording: Data Storage Systems Center (ERC) • Center for Extreme Ultraviolet Science and Technology (ERC)
Industrial, Manufacturing & Operational Systems	• Additive manufacturing (selective laser sintering, 3D printing) • "Geometric engine" at the core of 3D CAD • Automatic generation of machining instructions from 3D CAD models • Early math that allows the efficient scheduling algorithms for logistics of supply chains and airlines • Packaging Research Center (ERC)
Mechanical	• Center for Computer Integrated Surgical Systems and Technology (ERC) • Center on Mid-InfraRed Technologies for Health and the Environment (ERC)
Education and Workforce Development	• Recruitment of engineers

THE COMMITTEE'S FRAMEWORK FOR IDENTIFYING ENGINEERING IMPACTS ON SOCIETY

The committee was tasked to "identify up to 10 extraordinary engineering impacts made possible by NSF investments from 1950 onward" and was directed that

> [t]hese impacts might include expanded technological and social capabilities, scientific breakthroughs, and improvements in economic opportunity. They could have led to improvements in individual quality of life, national security, population health, manufacturing services, infrastructure resilience, and public policy, among others.

The framework the committee established to fulfill this task was informed by both this guidance and other considerations. Primary among these considerations was that the committee determined that it would not single out a "top 10" set of achievements. As this report makes clear, NSF support of engineering education, research, careers, and institutions has resulted in numerous innovations that have had and continue to exert profound societal effects. It is the committee's opinion that any attempt to distinguish the best 10 of these—however "best" might be defined—would necessarily exclude numerous significant achievements and obscure the breadth and magnitude of the impacts that NSF has had.

Instead, akin to efforts like the Nifty 50 and Sensational 60, the committee chose to highlight 10 *exemplary* impacts that had clear connections to NSF support; would be easily understood and appreciated by a general audience, and that covered a broad spectrum of characteristics:

- the time period specified in the statement of task (1950 onward)
- engineering disciplines—biomedical, civil, computer, electrical, …;
- the forms of impact—economic, social, technological, …;
- the nature of the support provided—educational, career, basic research, applied research;
- the recipients of the support—individuals, centers, academic institutions, private-sector organizations, partnerships, ….

Details, including citations to relevant literature and in some cases NSF grants, are included to document the reasoning behind the choices made.

In keeping with the statement of task directive to "engage young people from all segments of society" the committee chose impacts that lend themselves well to storytelling. The descriptions feature stories intended to draw the attention of and inspire diverse audiences, and place special emphasis on highlighting the work of researchers from historically underrepresented groups in engineering.[30] The text is intended to enhance the narrative describing the impact rather than detail the history of the underlying technology or program. While it does not include every achievement or individual who has made a significant contribution, the committee acknowledges the role that they played in bringing these accomplishments about.

The committee's decision making was informed by its own knowledge and experience along with staff-conducted research on candidate impacts and the resources identified above: the NAE member questionnaire responses, input received from NSF staff, NSF's own lists of its

---

[30] The committee followed the framework set out by NSF's biannual *Diversity and STEM* report (NSF, 2023b), which defines underrepresented groups as those including women, minoritized racial groups (African Americans, Hispanics or Latinos, and American Indians or Alaska Natives), and people with disabilities.

greatest achievements and lists generated by others, and the people who presented at the committee's symposium and the impacts that they identified.

In making their decisions about what constituted an impact "brought about" by NSF investments, the committee took the same expansive view of this question as the agency did in the *Nifty 50*, *Sensational 60*, and History Wall compilations listed above. NSF played a role in the achievements noted in these compilations, but the nature of that role varied. In some, it had a primary or even seminal part in the technology or program being highlighted while in others it carried on or enhanced work that had originated elsewhere, funded the development of a crucial piece of a larger "puzzle", or supplied early support for a researcher who went on to make significant contributions. All of these, in the committee's perspective, could constitute impacts worthy of inclusion in its list.

The committee was also mindful that engineering research and education are wide-ranging activities and that federal support makes up only a fraction of the total funding devoted to that work, with the private sector providing backing in some circumstances, and state and local governments, academic institutions, nonprofit organizations and others subsidizing some initiatives. Further, within the federal sphere, there are a number of agencies that support engineering-related research—prominently, the Department of Defense (DoD), including the Defense Advanced Research Projects Agency (DARPA), the Department of Energy, National Institutes of Health (NIH), National Aeronautics and Space Administration (NASA), and the National Institute of Standards and Technology (NIST). This is especially true for impacts in fields like computer science and engineering, where other organizations may have been the primary funding source, but where NSF played a significant role in bringing about particular impacts. These other sources of support are acknowledged where appropriate.

It is also true that NSF does not play a prominent part in supporting research in some engineering disciplines, resulting in their not being cited explicitly on the committee's list even though these fields are responsible for innovations that have had material societal impacts. The detailed accounts presented in this report do note, though, contributions in many of these fields.

EXEMPLARY ENGINEERING IMPACTS ON SOCIETY
IDENTIFIED BY THE COMMITTEE

The 10 exemplary engineering impacts on society brought about by NSF investments that were identified by the committee run the gamut from specific technologies to areas of research to programs that provide support. They are, in alphabetical order,

1. additive manufacturing (and, in particular, 3D printing)
2. artificial intelligence
3. biomedical engineering (and, in particular, rehabilitative engineering advancements)
4. cybersecurity
5. engineering education and early career support
6. materials science and engineering
7. NSF Centers (Engineering Research Centers and others)
8. NSF contributions to internet advancements
9. semiconductors and integrated circuits
10. wind energy technology

The programmatic areas—"NSF Centers" and "engineering education and early career development"—qualified as exemplary impacts in the committee's view because of their scope, which touches on virtually all engineering disciplines; their cumulative economic effect; the number of persons they have affected; and, of course, the range of breakthroughs they made possible through the support of research efforts and the people who achieved them

The descriptions of these impacts presented in the sections below describe:

- the technology, advancement, or program;
- the history and nature of NSF's support, acknowledging the role of other organizations that were involved in bringing about the impact;
- examples of specific technologies, outcomes, or other forms of impact resulting from the support;
- the ways in which the impact has affected society; and
- stories of one or more NSF-funded people involved in the impact.

They include references documenting the research initiatives, programs, funding, technologies, and people cited.

ADDITIVE MANUFACTURING

The emergence of additive manufacturing, particularly three-dimensional (3D) printing, has transformed traditional manufacturing by enabling efficient, digitally guided layer-by-layer construction of objects, offering enhanced material usage, design flexibility, accelerated production, and precise output compared to conventional techniques. While widely embraced by students and hobbyists in homes, classrooms, and maker spaces[31] around the country, 3D printing finds significant application in aerospace, automotive, medical, and other industries globally. Its versatility spans from prosthetics, dental implants, and artificial organs to artificial reefs, personal protective equipment, and food production (Agarwal, 2022; Heimgartner, 2022; Nelson, 2023; Sheela et al., 2021). A recent analysis valued the 2022 global additive manufacturing market at $14.5 billion (Vantage Market Research, 2024).

In 2013, NSF Assistant Director for Engineering Pramod Khargonekar praised additive manufacturing for "chang[ing] the way we think about the manufacturing process . . . by reduc[ing] the time, cost, and equipment and infrastructure needs that once prevented individuals and small businesses from creating truly customized items and accelerat[ing] the speed at which new products can be brought to market" (NSF, 2013a). In May 2022, the Biden Administration in collaboration with major U.S. manufacturers launched "AM Forward," aimed at bolstering the adoption of 3D printing among small and medium-sized manufacturers for supply chain resilience (CEA, 2022). This initiative seeks to use 3D printing's capabilities to drastically reduce lead times, material costs, and energy consumption compared with traditional processes (ASTRO America, 2022). This discussion delves into the history of additive manufacturing and its multifaceted impacts on society, examining the engineers behind many of these steps forward and how they have reshaped traditional manufacturing paradigms.

---

[31] Collaborative workspaces with equipment such as 3D printers.

History

The development of additive manufacturing and 3D printing is a story braided together by commercial, academic, and federal support and research. In the 1960s, the now defunct Teletype Corporation revolutionized printing with the invention of inkjet technology which enabled precise printing of ink up to 120 characters per second. This breakthrough not only facilitated rapid printing but also laid the foundation for the widespread adoption of consumer desktop printing. Building on this success, Teletype experimented with the use of melted wax instead of traditional ink, which led to Johannes F. Gottwald's 1971 patent. By solidifying liquified metal into predetermined shapes layer by layer, this prototype heralded a new era of rapid prototyping and additive manufacturing.

In the 1980s, a flurry of patents was filed for different types of 3D printing technologies, yet many were disregarded for seemingly having no commercial appeal. Raytheon filed a patent in 1982 to use powdered metal to add layers to an object. In 1984 entrepreneur Bill Masters filed a patent for a process called Computer Automated Manufacturing Process and System, which mentioned the term 3D printing for the first time. Another 1984 patent from France described additive manufacturing using stereolithography (explained below) but was also disregarded for a lack of "business perspective."

In 1984, Chuck Hull, an American inventor, helped develop and coined the term "stereolithography" (SLA) and subsequently founded the company 3D Systems. Based on Hull's previous patent for curing photopolymers (materials that undergo a chemical change when exposed to light) using radiation, his design sent spatial data from a digital file to the extruder of a 3D printer to build up the object one layer at a time, hardening each with light. It led to the release of the first-ever 3D printer, the SLA-1, in 1987, marking the dawn of a revolutionary era in manufacturing and design. Following Hull's innovation, the late 1980s and early 1990s witnessed the emergence of various other 3D printing technologies creatively using distinct materials and methodologies such as Fused Deposition Modeling, Selective Laser Sintering, and PolyJet (NSF, 2013d). The NSF Engineering Directorate took note of these developments and established the Strategic Manufacturing (STRATMAN) Initiative, led by the Division of Civil, Mechanical, and Manufacturing Innovation. STRATMAN provided support for research in the additive manufacturing field during this pivotal period of the late 1980s and early 1990s which saw the advent of its foundational technologies (Weber et al., 2013). Between 1986 and 2012, NSF made 593 additive manufacturing awards.

With 3,822 additive manufacturing patents filed from 1975 to 2011, the United States has been home to several of the most successful additive manufacturing companies, including 3D Systems, Stratasys, Z Corporation, and Solidscape (Wohlers and Gornet, 2015). Innovation in the field has become dominated by the private sector, especially when it comes to the total number of patents and the continual advancement of the technology beyond initial discovery. That said, many important and foundational milestones in additive manufacturing can be traced to federal funding, including support from NSF and DARPA (Table 4-5; Weber et al., 2013).

As an example of a company derived from academic breakthroughs, consider the case of Emanuel "Ely" Sachs[32]. Sachs became a faculty member at MIT in mechanical engineering in 1986 and worked from 1988 to 2002 on rapid prototyping and 3D printing, one of the pioneers of the "binder jet" method in 1989. Sachs and his team developed this method, which involves depositing a binding agent onto a layer of powder material, selectively bonding the powder

---

[32] Elected to the NAE in 2016.

particles together in the shape defined by the digital design. In 2015, Sachs co-founded Desktop Metal, a company that specializes in providing innovative 3D printing solutions for metal manufacturing processes (Desktop Metal, n.d.).

TABLE 4-5 Foundational and NSF-Influenced Advanced Manufacturing Patents and Processes

Advanced Manufacturing Process	U.S. Patent Number and Title	Inventor(s)	Application Year
Vat photopolymerization	4575330: Apparatus for production of three-dimensional objects by stereolithography	Charles Hull	1984
Powder bed infusion	4863538: Method and apparatus for producing ports by selective sintering	Carl Deckard	1986
Material extrusion	5121329: Apparatus and method for creating three-dimensional objects	S. Scott Crump	1989
Binder jetting	5204055: Three-dimensional printing techniques	Emanuel Sachs John Haggerty Michael Cima Paul Williams	1989
Sheet lamination	4752352: Apparatus and method for forming on integral object from laminations	Michael Feygin	1987
Contour crafting	5529471: Additive fabrication apparatus and method	Behrokh Khoshnevis	1995

SOURCE: Adapted from: Peña et al. (2014). Reprinted with the permission of the publisher, Mary Ann Liebert, Inc.

Numerous influential minds in additive manufacturing have propelled the field forward, such as Joseph Beaman, the Earnest F. Gloyna Regents Chair in Engineering at the University of Texas at Austin (UT). He is best known for his work on "Solid Freeform Fabrication" (SFF), a foundational technique which allows intricate solid objects to be created directly from a computer model without the need for custom tooling or specialized expertise. Beaman was the first academic to begin exploring this field in 1985 and coined the term in 1987 (TAMEST, 2013; UT, n.d.). In 1984, he was an inaugural winner of NSF's Presidential Young Investigator Award[33] to support his research, and he has subsequently received nearly $3.2 million in total NSF funding. Among the notable advancements originating in Professor Beaman's laboratory was Selective Laser Sintering (SLS), pioneered by Beaman and his student, the late Carl Deckard.

Beaman recounted how Deckard's grades weren't the best, but he advocated for Deckard's advisor to "let this guy in [to the graduate program], I think he's got some potential" (Davies, 2020). Together, they developed and successfully commercialized SLS, a process wherein a laser is used to melt and fuse together a powder to create a solid form, layer by layer. Deckard came up with the idea of SLS as an undergraduate, and as a graduate student under Beaman, the two of them secured a $30,000 NSF grant[34] to further develop the technology and construct a prototype machine. UT licensed SLS technology—a university-first—and it was spun off to a commercial operation in 1987 (TAMEST, 2013).

---

[33] Award # 8352272.
[34] Award # 8707871.

Jennifer Lewis is the Hansjörg Wyss Professor of Biologically Inspired Engineering at Harvard University. She is best known for her work in building the world's first 3D printed battery in 2013 (Wyss Institute, 2013). In the early 2000s, she 3D-printed scaffolds that mimicked the natural structure of bones and tissue in the body to induce natural bone growth and heal wounds using ceramic based materials (Jackson, 2017). Her lab develops "inks" with functional properties: cell-laden ones to print 3D tissues, or conductive inks that flow through rollerball pens at room temperature to draw functional circuits on paper. Lewis works with high-school teachers to incorporate these inexpensive pen-on-paper electronics in their classes so students can explore electrical engineering through circuit design. With over 15 NSF awards, including a Young Investigator[35] and CAREER Award[36] in 1994, Lewis has translated academic research into commercial ventures by co-founding start-up companies such as Voxel8 (acquired by Kornit), which manufactures multi-material 3D printing technology, and Electroinks Inc., which produces a reactive silver ink used in printed electronics. She developed a bioprinting platform for fabricating 3D human organ-on-chip models, which could eliminate the use of animal testing by the pharmaceutical and cosmetic industries and is a pioneer in 3D printed electronics, optical and structural metamaterials, soft robotics, and 3D vascularized tissues and organs.

Behrokh Khoshnev is known for developed contour crafting—a form of 3D printing using a computer-controlled crane to construct homes rapidly and efficiently with less manual labor—with NSF funding.[37] A later grant[38] helped Khoshnevis adapt the technology commercially and create a startup called Contour Crafting Corporation. This company specializes in rapid home construction as a way to rebuild destroyed homes after natural disasters, like the devastating earthquakes that have plagued his native country of Iran. It is now being considered for application in the construction of bases on Mars and Moon (CC Corp, n.d.). Khoshnevis also developed other powder-based additive manufacturing methods including Selective Separation Shaping, which works in zero-gravity conditions and can be used in space for the fabrication of spare parts and tools.

Additive manufacturing and 3D printing have revolutionized the way society advances and builds itself, with applications across practically every field of human endeavor. Its impact, from an economic, environmental, and medical perspective is already staggering, and its potential impact as the technology continues to evolve is almost impossible to predict. NSF, and particularly its Engineering Directorate, have played a critical role in the origin, growth, development, and advancement of the field of additive manufacturing and its associated technologies.

ARTIFICIAL INTELLIGENCE

The rapid advancement of artificial intelligence (AI) has brought about technological transformations that are reshaping our world in ways previously only seen in science fiction. From virtual personal assistants such as Siri, Alexa, and ChatGPT, to recommendation algorithms that suggest our next binge-worthy show on streaming platforms, to self-driving cars navigating our streets, AI has become an integral part of the 21st-century experience. Its impact

---

[35] Award # 9457957.
[36] Award # 9453446.
[37] Award # 9634962.
[38] Award # 0230398.

extends far beyond convenience and entertainment, touching upon fields as diverse as healthcare, finance, transportation, and education, with the promise of enhancing efficiency, solving complex problems, and opening new frontiers of innovation. That said, AI researchers are grappling with challenges like privacy issues, job displacement, and the imperative for diversity and inclusivity in AI development. AI can exhibit tendencies towards inherent biases and susceptibility to manipulation for malicious purposes, underscoring the intricate ethical dilemmas surrounding their deployment. These issues require careful stewardship from private industry, academia, and federal organizations as the U.S. navigates the transformative journey into the age of AI. This section explores some of innovations surrounding AI and complementary technologies, the people behind them, and the educational and research support that made their work possible.

Background

AI is "an umbrella term that means the use of computers to perform tasks that typically require objective reasoning and understanding" (Thomason, 2020). It refers to the longstanding research field, and is also used colloquially for contemporary applications that have become part of our everyday lives. Sub-areas of the broad field of AI include machine learning (ML) models like large-language models (LLM) and neural networks that are designed to handle text, images, and other data types. (IBM, 2023).

American AI research traces back to a pivotal workshop held at Dartmouth College in 1956. The attendees and their students would go on to shape the landscape of AI research throughout the 1960s, emerging as leaders in the field who achieved remarkable feats. Their computer programs broke new ground by mastering checkers strategies, solving algebraic word problems, proving intricate logical theorems, and even communicating in English (Crevier, 1993, pp. 52–54; McCarthy et al., 2006).

In the 1960s, investments in computing infrastructure supported the advancement of AI. The number of academic computer facilities grew from 100 in 1961 to over 2,000 by 1969, partially thanks to NSF's Computer Center Facilities program which awarded 414 infrastructure grants between 1959 and 1971 to various types of institutions such as research universities, liberal arts colleges, junior colleges, and even high schools (NSF, 2000d).

In recent years, advances in high-performance computing and software engineering have led to a burgeoning of applications of AI in everyday life, leading a 2024 International Monetary Fund report to observe that the field "is set to profoundly change the global economy, with some commentators seeing it as akin to a new industrial revolution" and concurrent effects on the labor market (Cazzaniga et al., 2024; p. 2). A different analysis estimated that AI could contribute over $15 trillion to the global economy by 2030 (pwc, n.d.). This section will explore just some of these impacts.

NSF's contributions to AI -related research is evidenced by the funding it has provided over the years. A 2024 analysis identified a total of $7.7B in AI-related grants and noted that "93 universities across the US have received AI-related NSF grants exceeding $20 million each, and 145 have received more than $10 million each" (Guerini, 2024). These include $225M to Carnegie-Mellon University, $219M to the University of California San Diego, $196M to the University of Illinois Urbana-Champagne, and $183M to the Massachusetts Institute of Technology.

Key Innovations and NSF's Contributions

Image Recognition

In 2006, Fei-Fei Li hypothesized that computers might learn like children by observing various objects and scenes (Hempel, 2018). Supported by an NSF CAREER Award, Li and her team created ImageNet, a database initially comprising 3.2 million hand-labeled images across 5,247 categories (Deng et al., 2009). ImageNet soon evolved into the ImageNet Large Scale Visual Recognition Challenge (ILSVRC) to see which algorithm could most accurately identify new images after being trained on this dataset (Russakovsky et al., 2015). Over seven years of ILSVRC, the database grew to over 14 million images across 20,000 categories, and object classification accuracy soared from 71.8 percent to 97.3 percent, surpassing human performance and igniting the current AI boom. In 2012, University of Toronto's ILSVRC winners, led by Professor Geoffrey Hinton (known as the "Godfather" of AI), introduced AlexNet (named after Hinton's Ph.D. student, Alex Krizhevsky, who designed it), which used deep neural networks to surpass the next-best recognition method by 41 percent (Gershgorn, 2017). AlexNet remains seminal in research architecture today. With NSF support,[39] Hinton had been exploring artificial neural networks since the 1980s. Meanwhile, others such as Yann LeCun were implementing neural networks in practical applications such as ATM check readers with the support of AT&T Bell Labs, and later, with NSF backing.[40] The contributions of Li, Hinton, LeCun, Yoshua Bengio and many others in deep learning and in AI image recognition laid the groundwork for modern tasks such as Facebook's photo tagging and self-driving car object detection. ImageNet taught us that the dataset on which algorithms are trained is as paramount for AI as the algorithms themselves.

Recommender Algorithms

In the early 1990s, recommender systems emerged such as John Riedl and Paul Resnick's NSF-supported GroupLens,[41] which collected news article ratings from readers to predict how much other readers would like an article before they read it. Journalist Malcolm Gladwell quoted Riedl's explanation of their system: "What you tell us about what you like is far more predictive of what you will like in the future than anything else we've tried" (Gladwell, 1999). With further NSF backing,[42] Riedl and Resnick founded the GroupLens lab, which was the first to train students in automated recommender systems, and also established commercial ventures like Net Perceptions. By the 2000s, e-commerce giants like Amazon and BestBuy had integrated recommender systems. From 2006 to 2009, Netflix sponsored a $1 million competition to enhance its recommendation algorithm (Bennett et al., 2007). The advent of AI techniques like neural networks and deep learning revolutionized recommender systems, enhancing accuracy and personalization.

Today, personalized recommendations are omnipresent, shaping our digital lives and influencing what people watch, buy, read, and listen to. However, Riedl and others were conscious that building effective systems necessitates not just advanced algorithms but also an understanding of user privacy, ethics, and diversity. In a 2012 talk on diversity issues on platforms like Wikipedia, Riedl urged students to consider "how to redesign socio-technical

[39] Award # 8520359.
[40] Award # 0535166.
[41] Awards # 9208546 and #9408708.
[42] Award # 9613960.

communities so that they work differently," for instance, "what would be a Wikipedia that was more welcoming, that worked better for women?" (Taraborelli, 2013). Shortly after this talk, Riedl succumbed to a years-long battle with cancer in 2013, leaving a legacy of impactful AI research and inspiring a generation of computer and social scientists to think of software design as a way to build better social systems (Taraborelli, 2013).

Speech Recognition

Speech recognition technology originated with Bell Labs' Audrey system in the 1950s which could understand the digits 0–9, and IBM's Shoebox machine in the 1960s, which recognized 16 English words (IBM, 2024). Early speech recognition systems faced numerous limitations, such as decreasing accuracy with different speakers, the need for pauses between words, and an inability to understand continuous speech. Dr. Raj Reddy at Carnegie Mellon University (CMU) and his team, supported by the Defense Advanced Research Projects Agency (DARPA) and NSF, made significant advances with his students developing Harpy in the early 1970s, which was capable of understanding over 1,000 words (Huang et al., 2014). Concurrently, spouses James and Janet Baker, also Reddy's students, took a "heretical and radical" approach to speech recognition, relying on a combination of phonetics and probability—how statistically likely it was that certain words would be paired together (Garfinkel, 1998). An article from MIT's *Technology Review* stated that "their system had no knowledge of English grammar, no knowledge base, no rule-based expert system, no intelligence. Nothing but numbers," yet, it soon outperformed other approaches (Garfinkel, 1998). The Bakers founded Dragon Systems in 1982, releasing such software as DragonDictate in 1990, which gained popularity among users with disabilities, and Dragon NaturallySpeaking in 1993, which solidified Dragon Systems as a leader in the field of continuous speech recognition (Garfinkel, 1998). The technologies pioneered by companies like Dragon Systems laid the groundwork for subsequent innovations like Google's Voice Search, Apple's Siri, and smart home devices. These advances, driven by AI and machine learning, have made speech recognition integral to modern life, with ongoing efforts focused on enhancing accessibility and adapting to evolving language dynamics.

Machine Learning

Machine learning plays a crucial role in advancing technologies related to AI. Although once considered synonymous with AI due to its learning and decision-making capabilities, it is more properly thought of as a subset of the field, developing alongside it until the late 1970s when it started to evolve independently. Today, machine learning is a vital tool that is utilized in numerous technologies, including several discussed in this and other sections of the report.

NSF's involvement with machine learning traces back to the late 1960's, when it funded studies that examined how computers could be used for pattern recognition. Later, it supported research on neural networks and large language models, which are used in generative AI applications like chatbots. It has since significantly expanded and the agency's current National Artificial Intelligence Research Institutes initiative, including programs that support the building of intelligent learning systems for STEM education (The INVITE Institute[43]); combine meteorological, soil, and crop yield data to identify climate-smart agricultural practices and boost

---

[43] Inclusive Intelligent Technologies for Education Institute – https://invite.illinois.edu/.

rural economies (AI-CLIMATE[44]); and apply machine learning to develop and operate safe, reliable, and ethical dynamic systems (Institute in Dynamic Systems[45]). NSF's partners in this effort, which is administered by the agency's Directorate for Computer and Information Science and Engineering, include the National Institute of Standards and Technology; the Departments of Agriculture, Defense, Education, and Homeland Security; and IBM Corporation.

NSF also supports foundational research in machine learning, notably that conducted by Tomaso Poggio, an MIT professor whose many titles include serving as director of the Center for Brains, Minds, and Machines,[46] an NSF center that seeks to understand how the human brain generates intelligent behavior and how that intelligence might be replicated by machines. He is considered to be one of the founders of the field of computational neuroscience and he and the Center have made key contributions to the biophysics of computation and learning theory, and developed an influential model of the processing of information by the visual cortex. Several of Poggio's students have gone on to make their own contributions in the field, including Amnon Shashua, who founded OrCam[47]—a start-up that uses AI technology to help blind and visually impaired people understand the environment around them via devices that can clip onto a pair of glasses.

Facial Recognition/Affect Recognition

The journey of facial recognition technology reflects both engineering progress and ethical dilemmas. Facial recognition technology, spurred by federal support, has evolved since the 1960s. Early efforts by Woody Bledsoe and his team manually plotted the coordinate locations of facial features—mouth, nose, eyes, and even hairline—and compared new photographs against this database to identify individuals with the closest numerical resemblance (Raviv, 2020). Much of his work was backed by the Central Intelligence Agency (CIA). The 1990s witnessed a groundbreaking advance with the development of the Eigenface method by Matthew Turk and Alex Pentland, partially funded by NSF,[48] which efficiently represents faces using linear algebra and serves as the foundation for numerous modern facial recognition algorithms. In 2001, Paul Viola and Michael Jones from MIT pioneered real-time face detection in video footage through their integration of the Viola–Jones object detection framework (Viola and Jones, 2004), resulting in AdaBoost, the first real-time frontal-view face detector, supported in part by Viola's NSF CAREER Award.[49] Some of the earliest clients of facial recognition technologies were the Department of Motor Vehicles (DMVs) to prevent people from obtaining multiple driving licenses, U.S. prison systems for automated identification systems, and law enforcement to track criminals across states (Gates, 2011). Meanwhile, Rosalind Picard and Rana El Kaliouby, supported by NSF grants,[50] sought to imbue AI with emotional intelligence by recognizing and identifying facial expressions. Their company, Affectiva, aimed to aid those on the autism spectrum but faced a moral dilemma when offered funding for surveillance purposes. They stood by their core values, refusing a $40 million offer and instead finding investors aligned with their vision. In a *Forbes* interview, El Kaliouby emphasized, "It was empowering

---

[44] AI Research Institute for Climate-Land Interactions, Mitigation, Adaptation, Tradeoffs and Economy – https://cse.umn.edu/aiclimate.
[45] https://dynamicsai.org/.
[46] https://cbmm.mit.edu/.
[47] https://www.orcam.com/.
[48] Award # 8719920.
[49] Award # 9875866.
[50] Awards # 0705647 and # 0087768.

because it demonstrated that you can navigate your path according to your principles" (Cohn, 2020). AI facial recognition remains controversial and has been restricted in some U.S. cities. These technological developments continue to spark debates on balancing security and individual rights and weighing the societal implications of AI progress.

Robotics

Since the beginning of AI, researchers have explored ways to integrate AI with robotics. In the mid-20th century, early experiments focused on creating machines that could mimic basic human tasks, such as navigating simple mazes or performing specific, pre-programmed actions. Early experiments laid the foundation for AI-driven robots capable of basic tasks. The 1980s and 1990s marked a significant period of advancement, with the introduction of more sophisticated algorithms and computational techniques. During this era, robotics research began to incorporate elements of machine learning, computer vision, and natural language processing, enabling robots to learn from experience, recognize objects, and understand spoken commands. This period also saw the establishment of dedicated research institutions and university programs across the United States, supported by federal agencies such as NSF, DARPA, and NASA. Some notable examples of AI robots and their innovators include:

- Alice Agogino is a pioneering figure in engineering education and AI who established the University of California, Berkeley's first integrated artificial intelligence hardware lab in the mid-1980s and received the NSF Presidential Young Investigator Award. Her long career at Berkeley has taken many turns, including founding a spin-off company, Squishy Robotics, which makes robots capable of surviving drops of up to 1,000 feet for various applications, including disaster response, military use, the Industrial Internet of Things, and package delivery (Squishy Robotics, 2024).
- Manuela Veloso, a prominent figure in AI and robotics, has spearheaded research at Carnegie–Mellon University in collaborative robots, or *cobots*, aiming to enhance productivity and efficiency by combining machine learning with autonomous decision making. With an NSF CAREER award, her work focuses on creating autonomous agents that integrate cognition, perception, and action to address various tasks such as navigating indoor environments and participating in team-based activities, reflecting a future where humans and machines collaborate more closely (Brandom, 2016).
- Dr. Yulun Wang, dubbed the "father of modern surgical robotics," (Southern California Biomedical Council, 2020) founded InTouch Health, now part of Teladoc Health, revolutionizing healthcare delivery by enabling remote surgical procedures and consultations through his NSF-funded innovations in telepresence and surgical robotics, with a strong emphasis on humanitarian efforts to provide healthcare in conflict-affected regions such as Ukraine.
- Sebastian Thrun, renowned for his pivotal role in developing self-driving cars as a Stanford University professor and Google's self-driving car project founder, achieved significant milestones in AI-driven transportation, with his research in robotics, artificial intelligence, and machine learning having been supported by NSF grants. Additionally, his initiative to democratize education through Udacity,

offering free online courses to millions globally, reflects his commitment to inclusive and accessible learning opportunities (Hill, 2023).

Observations

The rapid evolution of AI has brought about transformative changes across various aspects of society, from health care to entertainment and education, but also brings with it ethical, privacy, and job displacement concerns. This revolution—which traces back to pivotal early investments in people and technologies by federal agencies such as NSF—has led to groundbreaking innovations in AI-driven fields such as image recognition, recommender algorithms, speech recognition, facial recognition, and robotics. Exemplary researchers and engineers have not only advanced the capabilities of AI systems but also grappled with ethical considerations, privacy concerns, and the need for diversity and inclusivity in AI development. Through efforts such as making "broader impacts" one of two criteria used to evaluate requests for support, NSF has required researchers to take a well-rounded approach to their work. The National Artificial Intelligence Research Institutes (AI Institutes) program, initiated by the NSF's CISE directorate, is a significant collaborative effort among federal agencies and private organizations to advance AI research, education, and workforce development, with 25 institutes established across 40 states and D.C., and will play a key role in implementing the 2023 Executive Order on safe and trustworthy AI.[51] As AI becomes increasingly integrated into various sectors, the collective effort of academia, industry, and federal agencies will be crucial in navigating the ethical, legal, and societal implications of AI technologies, while also ensuring equitable access and fostering innovations that yield benefits to society.

BIOMEDICAL/REHABILITATION ENGINEERING

If you or anyone you know has ever gotten an MRI, had genetic analysis, or uses prostheses, you have biomedical engineering to thank. Engineers play an essential role by working cross disciplinarily to help design the technology, delivery methods, diagnostic tools, and devices that are advancing the horizons of modern medicine and having direct impacts on improving and saving peoples' lives all over the globe. For the past 60 years, NSF, amongst other federal agencies like the National Institutes of Health, Department of Veterans Affairs, Department of Defense, and private organizations, has played an instrumental role in funding this research and supporting the education of the professionals who have made these breakthroughs (Biomedical Engineering Society, 2004).

Areas of biomedical engineering supported by NSF include but are not limited to:

- Biophotonics: the use of light-based technologies to visualize and analyze cells and tissue.
- Biosensing: harnessing biological molecules to measure the presence of various substances.
- Cellular and biochemical engineering: the manipulation and optimization of cells and biochemical pathways to develop products and processes for medical, industrial, and environmental applications (NSF, 2023c).

---

[51] The description of the AI Institutes program in this chapter contains additional information on this issue.

Another prominent area of advancement is rehabilitation engineering, the "the use of engineering principles to 1) develop technological solutions and devices to assist individuals with disabilities and 2) aid the recovery of physical and cognitive functions lost because of disease or injury" (National Institute of Biomedical Imaging and Bioengineering, 2016). There are an estimated 1.85 billion people with disabilities globally, a number greater than the population of China (Donovan, 2020) and the demand for assistive/rehabilitation devices will only grow in the coming years owing to the aging of the general population.

Several programs at the NSF support rehabilitative research, including the Disability and Rehabilitation (DARE) program (started in 2017); Biomechanics and Mechanobiology (started in 2016); Engineering of Biomedical Systems (started in 2015); Smart and Connected Health (started in 2013); National Robotics Initiative (subfocus on assistive robotic technology, started in 2011); Human Centered Computing (subfocus on assistive and adaptive technology, started in 2013);and the Mind, Machine, and Motor Nexus program (started in 2018) (NSF, n.d.-h; NSF, 2013a; NSF, 2013b; NIH, 2011), which have collectively contributed over $292 million towards assistive and rehabilitative technological advances. In late 2022, for instance, NSF's Convergence Accelerators awarded $11.8 million to 16 teams to design projects enhancing the quality of life for people with disabilities (NSF, 2022d).

A 2022 NSF article (Zehnder and Kulwatno, 2022) highlighted some of the more recent assistive technological advances, including:

- A custom exoskeleton (a wearable device that supports or enhances movement)   to be worn over the leg that measures the changing mechanical properties of the knee during complex, real-world situations to enhance the functionality of future rehabilitation robotics and protheses.[52]
- Actuators (devices that convert energy into mechanical motion) made entirely of soft, biocompatible materials that can be safely implanted in the ear and deliver hearing aid electrodes without the rigidity of previous implants which can damage inner ear structures.[53]
- Organic actuators that can recreate the complex sensations of touch including roughness, adhesion, softness, and moisture in virtual reality settings for myriad applications such as virtual reality rehabilitation and simulated surgery.[54]
- Lighter assistive exoskeletons that help paralyzed people perform daily activities via the combination of functional electrical stimulation and assistive robotics technology.[55]
- Lab-grown neural tissues that can recognize and decode brain signals, which may eventually be able to replace damaged brain tissues and restore function to individuals who have had strokes.[56]

For decades, researchers and initiatives supported by NSF have been laying the foundation for these modern advances. The field of tissue engineering, for instance, was championed by Y.C. Fung of the University of California at San Diego, who coined the term at a

---

[52] Principal investigator (PI) Elliot Rouse, Award #1846969, University of Michigan.
[53] PI Jaeyoun Kim, Award #1605275, Iowa State University.
[54] PI Darren Lipomi, Award #1929748, University of California-San Diego.
[55] PI Eric Schearer, Award #2025142, Cleveland State University.
[56] PI An Hong Do, Award #2223559, University of California-Irvine.

1987 meeting sponsored by NSF (Viola et al., 2003). Soon thereafter, the discipline was defined at another NSF-sponsored workshop as "the application of principles and methods of engineering and life sciences toward fundamental understanding . . . and development of biological substitutes to restore, maintain and improve [human] tissue functions" (Skalak and Fox, 1998). Between 1987 and 2001, the NSF invested more than $70 million in the field (Viola et al., 2003).

The agency continues to support researchers and labs working in tissue engineering, such as Gilda Barabino. Dr. Barabino—Professor of Biomedical and Chemical Engineering and President of the Olin College of Engineering, as well as the 2022 President of the American Association for the Advancement of Science—was the first African American admitted to the chemical engineering program at Rice University and only the fifth African American woman in the United States to obtain a doctorate in chemical engineering (NASEM, 2023). Currently, her research focuses on developing novel, wavy-walled bioreactors that enhance mixing while reducing shear that can damage cells. This work was partially supported by the NSF Visiting Professorship for Women in Science and Engineering program.[57] Dr. Barabino has received over $2.5 million in NSF support throughout her career to not only conduct research but also to help broaden participation of minority engineering faculty, enhance interdisciplinary collaboration, and strengthen ties to enterprise.

Cutting-edge research on technologies that interface with neural systems has also led to previously unimaginable advances such as restoring sight to the blind. Retinal protheses, also known as "artificial retinas," "retinal chips," or "bionic eyes," are implantable electronic devices that have restored some vision and perception of light and motion to some retinitis pigmentosa patients by recreating vision by capturing images from photosensor arrays and sending corresponding electrical signals to nerve cells in the retina (Chamot, 2003). This can help people with diseases such as retinitis pigmentosa or age-related macular degeneration. This work was highlighted in the NSF's "Nifty 50" list of innovations in 2000, when "researchers [were] a few years from permanently implanting an eye chip into a blind person" (NSF, 2000c). By 2010, NSF's "Sensational 60" list described how "researchers have begun implanting retinal prosthesis in blind people, . . . allow[ing] patients who had not seen light to see light and to make out some shapes and sizes" (NSF, 2010b). As of early 2023, the number of electrodes in artificial retinas has risen from 16 in the early 2000's to 240 (NASEM, 2023). Patients have been able to recognize letters and even play basketball with their grandchildren, and researchers are working on cameras that can be implanted into the eye to transmit signals to the artificial retina, rather than being worn on external lenses. This research has been spearheaded by Dr. Mark Humayun, co-inventor of the Argus retinal prosthesis system, with over $42.6 million in NSF support since 1998, including a grant of over $37 million to establish the University of Southern California's Biomimetic Microelectronic Systems Engineering Research Center from 2003 to 2015 (currently the Ginsburg Institute for Biomedical Therapeutics) where many of these advances were made, as well another NSF grant of nearly $2 million in 2019 to continue this research. In the authoring committee's August 2022 information-gathering symposium (NASEM, 2023), Humayun described how "on a personal note, it was also of extreme interest to me because my grandmother, who raised me, went blind from diabetic retinopathy."

Engineers often have life experiences that motivate them to make a difference in society through their work. Direct experience also lends itself to improved designs (Henderson and Golden, 2015), underlining the significance of supporting researchers who themselves have disabilities in the development of inclusive and accessible assistive technologies. Dr. Rory A.

---

[57] Award # 8211920.

Cooper, for instance, is a U.S. Army veteran with a distinguished professorship at the University of Pittsburgh in the Department of Rehabilitation Science and Technology (University of Pittsburgh, n.d.-c). He incurred a spinal cord injury through his military service and uses a wheelchair. Since becoming disabled, he has earned all three of his degrees, completed marathons, and founded the university's Human Engineering Research Laboratory (HERL). The mission of HERL is to "continuously improve the mobility and function of people with disabilities through advanced engineering in clinical research and medical rehabilitation," and the lab celebrated its 30-year anniversary in 2023 (University of Pittsburgh, n.d.-a). Dr. Cooper stated that one of the NSF's best programs is the Research Experience for Undergraduates (REU) program (NASEM, 2023). He leads an REU program at HERL[58] entitled "American Student Placements in Rehabilitation Engineering (ASPIRE)" which promotes "greater involvement and understanding of Rehabilitation Engineering and assistive technology—while fostering an understanding of the problems faced by individuals with disabilities" (University of Pittsburgh, n.d.-b). Over a quarter of the past REU participants had reported an impairment that limits one or more daily activities. Examples of innovative projects coming out of the Pitt ASPIRE REU program include designing a waterproof wheelchair that increases accessibility to places like waterparks and expands play possibilities for children, applying functional MRI to understand how cortical hand control is affected by spinal cord injuries, and developing a custom mobile health application to promote physical activity for those who use manual wheelchairs. Dr. Cooper has also been supported by NSF programs such as the ICorps for Learning pilot program and the Experiential Learning for Veterans in Assistive Technology and Engineering program.

Another biomedical engineering researcher whose innovations are already helping society is Dr. Ayanna Howard, current Dean of The Ohio State University's College of Engineering. She believes that "every engineer has a responsibility to make the world a better place. We are gifted with an amazing power to take people's wishes and make them a reality" (Georgia Tech, n.d.-a). With nearly $7.5 million in funding from NSF to date, Dr. Howard has applied her extensive experience in robotics to develop cutting edge robots to, among a host of functions, assist in therapeutic activities and pediatric rehabilitation for children with special needs including Down syndrome, autism, and cerebral palsy. While working at the Georgia Institute of Technology (Georgia Tech) as director of the Human-Automation Systems Lab (HumAns) from 2005 to 2021 (Georgia Tech, n.d.-b), Howard received NSF support for projects such as "Accessible Robotic Programming for Students with Disabilities",[59] "Robot Movement for Patient Improvement" aimed to "to fuse play and rehabilitation" to help children with pediatric rehabilitative needs,[60] and "An Accessible Robotic Platform for Children with Disabilities" to scale-up and make this platform commercially viable.[61] In 2013, Howard channeled this research into a spinoff company Zyrobotics, which received nearly $900,000 in seed funding from NSF's Small Business Innovation Research (SBIR) program[62] (SBIR, n.d.). The examples of Howard and others demonstrate how federal funding for rehabilitation engineering research can lead to the creation of spin-off companies that bring innovative assistive device technologies to consumers and benefit the economy. In addition to her passion for inclusive technology, Howard has been involved in several initiatives to recruit and retain

---

[58] Award # 1852322.
[59] Award # 0940146.
[60] Award # 1208287.
[61] Award # 1413850.
[62] Awards # 1447682 and # 1555852.

minority groups in engineering, including serving as lead investigator for Georgia Tech's Summer Undergraduate Research in Engineering program in Robotics[63] from 2009 to 2021. Her book, *Sex, Race, and Robots: How to Be Human in the Age of AI* explores how human biases affect AI algorithms and the implications for society (Howard, 2020).

In September 2022, the Biden–Harris administration signed an executive order on "Advancing Biotechnology and Biomanufacturing Innovation for a Sustainable, Safe, and Secure American Bioeconomy" in which the NSF was tasked with "identifying high-priority fundamental and use-inspired basic research goals to advance biotechnology and biomanufacturing and to address the societal goals" (White House, 2022a). In response to this executive order, the NSF's 2022–2026 Strategic Plan (NSF, 2022a) emphasized the organization's aims to "expand our strategic leadership across emerging areas" (p. 3) including biotechnology, a field which "will advance the U.S. bioeconomy, accelerating the ability to harness biological systems to create goods and services that contribute to agriculture, health, security, manufacturing and climate resilience" (p. 23). The White House Office of Science and Technology Policy subsequently launched the National Bioeconomy Board to complement this effort (OSTP, 2024).

There are thus a wide range of biomedical engineering achievements made possible with NSF investments. The cited researchers and assistive technologies are illustrative of the agency's impact in this field. Although it has been over 33 years since the Americans with Disabilities Act and the Individuals with Disabilities Education Acts of 1990 were signed into law, disabled rights activists continue to fight for greater inclusion and accessibility to public spaces, equitable education, and opportunities. Support for progress in research areas like biomedical and rehabilitation engineering not only promises to enhance the quality of life for the millions of people living with disabilities, but also to bolster the influence of disabled and underrepresented researchers at the forefront of these fields.

CYBERSECURITY

The Information Age has ushered in an era marked by unprecedented connectivity and interdependence. From the 1960s through 1980s, advances such as computer time-sharing, local area networks, broader networks of the ARPANET and internet, semiconductors, and advances in storage transformed computing technology. In the 2000s and 2010s, technologies such as personal computers, the World Wide Web, smartphones, IoT (Internet of Things), and cloud computing became widespread, making digital technology a part of our daily lives. However, this interconnectedness has also exposed societies to novel vulnerabilities, with the realm of cyberspace emerging as a critical frontier.

The importance of cybersecurity in our modern world cannot be overstated, as it stands as a bulwark against malicious actors—from both criminals and nation-states— seeking to exploit digital infrastructure for nefarious ends. The exponential growth of digital networks has intertwined critical systems, economic services, and personal lives, making the safeguarding of sensitive information and the integrity of digital systems paramount. A breach in cybersecurity not only jeopardizes individual privacy but has the potential to cripple essential services, compromise national security, and weaken the foundations of a globalized society. In the United States, as with much of our technological advancement, cybersecurity progress has been a group effort among government agencies, private industry, and academia.

---

[63] Awards # 0851643, # 1263049, and # 1757401.

In the 1970s and 1980s, threats to computers shifted from the mere physical security of locks to the digital security of software systems. This need came about with the advent of time-sharing systems that linked minicomputers or terminals to larger mainframes, either locally or on pioneering commercial networks such as Tymshare's TYMNET. Pioneering research and development (R&D) efforts on digital access control technology and intrusion detection systems were sponsored by the Department of Defense (DoD) and the National Security Agency (NSA). More specifically, the DoD/NSA collaborative National Computer Security Center; the military branches (especially the Air Force); and Federal Funded Research and Development Corporations (especially MITRE, RAND, SRI, and System Development Corporation) did path-breaking work on standards and certification mechanisms. This R&D was important to protecting the nation's classified information and intelligence infrastructure but had relatively little impact on security tools and technology in the private sector, which was often far less robust. This was true for the early internet as well.

A largely different group of institutional and individual academic and government actors spawned the internet. Central to this were standards groups such as the Internet Engineering Working Group (IEWG), academic researchers, and the National Science Foundation (NSF). The NSF provided research and development to fund regional networks and facilitated the underlying internet networking backbone that became NSFnet.[64] These groups then worked at privatizing this infrastructure into today's internet in the first half of 1990s. The IEWG, NSF, and other government and academic and industry partners (MCI, IBM, etc.) prioritized interoperability and access.

Rapid deployment and interoperability in computer networking and the internet had many economic and social benefits, but it resulted in security vulnerabilities. Focusing events, such as the Morris Worm—generated by then Cornell graduate student Robert Tappan Morris—in 1988 which disabled at least a tenth of the internet for 72 hours (MIT, n.d.-b), made clear how vulnerable the Internet was to various types of malware, viruses, distributed denial of services, and other types of attacks and misuse (Karger et al., 2002; Spafford, 2003).

Recognizing this early on, NSF began to support computer security research in different parts of the agency, particularly by divisions of the Computer and Information Science and Engineering (CISE) Directorate, funding research at universities and other research organizations. Since the 1990s, NSF programs such as Secure and Trustworthy Cyberspace (SaTC) and its predecessors , as well as collaborations of CISE with the Engineering Directorate and other directorates and programs within NSF, have helped develop and advance significant technologies and policies in cybersecurity. This continued in the later 1990s and expanded before the end of the new millennium's first decade. In 2009, this resulted in NSF CISE's Trustworthy Computing Program, led by Carl Landwehr, a senior scientist in computer security who had worked at the Naval Research Lab for many years. In several years this effort expanded (in name, dollars, and a greater agency-stretching makeup) into NSF's Secure and Trustworthy Cyberspace Program. Centered in CISE and collaborating with other NSF Directorates— Engineering, Math and Physical Sciences; Social, Behavioral and Economic Sciences; and Education and Human Resources—SaTC and its partners have contributed greatly to cybersecurity and privacy.

Today's threats to cybersecurity and privacy are complex, extensive, and constantly accelerating and evolving. The offensive capabilities of adversarial state actors like Russia,

---

[64] Former NAE President, Bill (William) Wulf was a key driver of this transformation when he was Assistant Director of NSF's Directorate of Computer and Information Science and Engineering (CISE).

China, North Korea, and others are quite advanced, as are those of the United States. On the offensive side, the U.S. military branches (especially U.S. Cyber Command and the Air Force), intelligence agencies (NSA and the Central Intelligence Agency), DoD, and universities such as Naval Postgraduate School and Air University, tend to take the lead. While all of these entities extensively fund defensive research and development work–often intertwined to varying degrees–NSF and SaTC are particularly impactful on defense against a vast range of cyberattacks. These include well-known methods, crimes, and data abuses, but also zero-day attacks, or malware techniques that are unforeseen (0 days to prepare against). Fundamental to this is an interdisciplinary and integrative approach to both basic and applied research for which NSF and SaTC have no equal in supporting research, education, knowledge, and infrastructure to better ensure cybersecurity and privacy. NSF designed SaTC precisely for this purpose.

The Assistant Director of CISE from 2011 to 2014, Farnam Jahanian,[65] provided critical leadership and support for the early SaTC initiative. Jahanian, an Iranian-American, emigrated to the United States as a teenager. His many cybersecurity accomplishments include co-founding Arbor Networks, where he and his team devised means for service providers to safeguard networks from zero-day threats, distributed denial of service attacks, and other risks. These internet security solutions have been widely adopted by internet service providers, cloud service providers, wireless carriers, and numerous other networks globally. During Jahanian's tenure as director of the NSF Directorate for Computer and Information Science and Engineering, he not only helped to launch interdisciplinary research and education initiatives such as SaTC, but also public–private partnership programs such as U.S. Ignite and the I-Corps. In an interview, Jahanian explained that "as an immigrant, serving my country was such an honor. I am very proud of my academic record and the research I have done. And I am proud of the students I have mentored, and the impact I have had through entrepreneurship. But, besides my children, what I am most proud of is my service in the public sector" (Sewald, 2019).

As Jahanian recounted, "the aim was to support fundamental scientific advances and technologies to protect cyber systems from malicious behavior, while preserving privacy and promoting usability" (Freeman, et al., 2019, p. 158). Keith Marzullo, who formerly led the SaTC program, stressed how an integrative agency-wide approach was essential to SaTC from its conception. He stated that they brought in:

> . . . [c]yberinfrastructure, which . . . was a separate office . . . to understand the infrastructure aspects of cybersecurity as well as the need to protect our supercomputing capacity. We brought in Math and Physical Sciences, because there's a whole aspect to quantum computing and the deep math associated with that, and we brought in social, behavior[al], and economic scientists, because if you were to look at the strategic plan that was published in 2010 on cybersecurity by the Office of Science and Technology Policy, you'll see they called out for emphasizing economic incentives (Freeman et al., 2019, p. 158).

As such, the SaTC program was wide-ranging and a major investment for NSF in helping to advance cybersecurity, privacy, and national security. In 2016 it represented a $160 million dollar program across NSF. Rarely does a relatively new program grow to this size so quickly at the foundation. This fast ramp-up is indicative of the range of cybersecurity and privacy threats

---

[65] Dr. Jahanian is currently (2024) the President of Carnegie Mellon University.

and the importance of basic and applied interdisciplinary and inter-directorate and program research and development. This remains true today, as the current program emphasizes.

Achieving a truly secure cyberspace requires addressing both challenging scientific and engineering problems involving many components of a system and vulnerabilities that stem from human behaviors and choices (NSF, n.d.-i).

What SaTC (and the ramping of funding with Trustworthy Computing leading to it) has accomplished is wide and varied, and the most meaningful work has often extended from the integrated, interdisciplinary research and infrastructure it supports. This is because understanding behavior is so critical to effective cybersecurity and protections of privacy. What follows are several examples of such support and its impacts.

Security, Privacy, and Usability

One computer scientist who has done significant work at the intersection of human–computer interaction and security and privacy research and education is Carnegie Mellon University (CMU) Professor of Computer Science and Engineering and Public Policy, Lorrie Faith Cranor. After receiving a doctorate that focused on human–computer interaction and privacy, she worked on computer security at AT&T Labs before joining the faculty of Carnegie Mellon. Her early work in the intersection between privacy, security, and usability included the seminal monograph *Security and Usability* (Cranor and Garfinkel, 2005). Shortly after arriving at CMU, Cranor received a 5-year NSF Integrative Graduate Education and Research Traineeship (IGERT) grant in 2009 for $2.92 million, "Usable Privacy Security," on which Cranor was Principal Investigator and Program Director. It allowed her to start and grow a pioneering laboratory that she named CUPs for Cylab Usable Privacy and Security. The usability-focused event, her book, and research/education of the lab coalesced to have tremendous impact. The event became SOUPs, an annual Symposium on Usable Privacy and Security. SOUPs rapidly grew, from roughly 70 attendees at the original event, to become a major computer science symposium and arguably the most diverse event in cybersecurity and privacy. Cranor's pioneering IGERT was at the cusp of Trustworthy Computing, growing into SaTC, which would become ever more interdisciplinary and inter-directorate and programmatic at NSF. The IGERT that had facilitated the start and early growth of CUPS established CMU as leader for graduate education and research on usability.

Cranor and her team's work has had significant implications for the password requirements that are part of our interactions with every website that uses our personal data. Their analyses of password strength, user behavior, and user sentiment under different sets of password composition protocols (i.e., requirements for length, special characters, or numbers)— supported in part by the NSF—have resulted in improved policies that put less burden on users while enhancing the security of their information (IGERT, 2011).

TRUST Collaboration for Trustworthy Infrastructures

One major challenge to secure infrastructures is add-ons placed on systems not designed to be secure. As such, research and education into the design of trustworthy infrastructures is extremely important and is an area that SaTC—and prior security and privacy NSF-funding—has targeted to affect positively. The NSF has led efforts for the creation of multi-university collaborations to have in targeted areas on a large scale. One valuable center launched with NSF funding and receiving $40 million over a 10-year period is the Team for Research in Ubiquitous Secure Technology, or TRUST (NSF, 2017). S. Shankar Sastry, the longtime Dean of Engineering at the University of California (UC), Berkeley, was the principal investigator (PI).

Along with TRUST's home base at UC Berkeley were partnerships with Carnegie Mellon University, Cornell University, San Jose State University, Stanford University and Vanderbilt University. It had four targeted areas—Financial, Health, and Physical Infrastructures, and the science of security—with a number of the partnering schools working on each of the four areas. In these domains, the center focuses on technical, operational, legal, policy, and economic issues in the design, development, deployment, and maintenance of trustworthy systems. It used early NSF support to bring in tens of millions of dollars from a range of private foundations as well as a host of industry partners.

More recently, in 2022 the NSF announced a $25.4 million investment through SaTC. These funds were awarded to various projects, including nearly $6.4 million for North Carolina State University's secure software supply chain project, which aims to protect the process of software development and distribution from vulnerabilities that hackers could exploit. Additionally, over $4 million was awarded to the University of Florida for research on privacy for marginalized groups, addressing the specific challenges faced by vulnerable populations and developing technologies to safeguard their personal information. Indiana University received nearly $3 million for a project centered on secure computation using trusted hardware, which involves using specialized computer chip hardware to perform sensitive computations securely, ensuring data confidentiality (Errick, 2022).

Observations

NSF in conjunction with other governmental agencies, universities, and private industry, has played a pivotal role in shaping the landscape of cybersecurity, addressing the evolving challenges posed by the rapid advancement of digital technologies. From the early days of computer time-sharing and the ARPANET to the current era of smartphones, cloud computing, and IoT, NSF has fostered research and development to safeguard privacy, commerce, and critical infrastructures. Through programs such as Secure and Trustworthy Cyberspace, NSF has facilitated interdisciplinary collaborations, bringing together experts from computer science, engineering, mathematics, social sciences, and more. The SaTC program's comprehensive approach, spurred by the leadership of individuals such as Farnam Jahanian and Carl Landwehr, has not only addressed well-known cyber threats but has also anticipated emerging challenges, including zero-day attacks. The success stories highlighted underscore the impact of NSF's investments in research, education, and infrastructure.

ENGINEERING EDUCATION AND EARLY CAREER SUPPORT

The NSF plays an important role in shaping the future of engineering and science by supporting greater diversity, equity, and inclusion in engineering education as well as early career stages nationwide. To achieve these goals, the foundation has implemented various initiatives, including the Faculty Early Career Development Program (CAREER), which encourages applications from underrepresented groups (NSF, 2022c). Additionally, the Engineering Education Coalitions Program (1990-2005) went beyond the traditional research grant model to support university consortia aimed at improving engineering education and broadening participation (Borrego et al., 2007). NSF emphasizes support for researchers who not only conduct high-impact research but also engage underrepresented groups into their research and education and transfer their knowledge into products and services that benefit our society. These investments ensure that diverse voices and perspectives help drive technological and societal advancement. This section highlights NSF's efforts in transforming engineering culture, practices, and policies through educational and early career support of diverse and underrepresented groups in engineering.

Example NSF Programs and Their Impacts

NSF ADVANCE

NSF ADVANCE was initiated in 2001 to increase women's representation and advancement in academic science and engineering careers. In the first 20 years, NSF ADVANCE invested over $365 million and supported more than 200 universities across the United States (Gold et al., 2022; NSF, 2021a). The first call for proposals had three tracks. The fellowship and leadership tracks provided support for individual women, yet this focus on "fixing the women" does not address the broader environment and culture in which they operate. The institutional transformation (IT) track focused on the institution instead of the individual, representing a new model at NSF for broadening participation that has been part of ADVANCE since its inception; indeed, the lessons learned by participants from all tracks informed later program directions (DeAro et al., 2019; Morimoto et al., 2013). The current ADVANCE program includes IT, Adaptation, Partnership, and Catalyst tracks. ADVANCE'S focus on gender (and, later, on other identities that intersect with gender) and systemic barriers in academic STEM careers is essential to developing a more diverse STEM workforce.

ADVANCE not only promotes institutional transformation and systematic changes but also facilitates both basic and applied knowledge production on perspectives within and about gendered academic organizations. In other words, ADVANCE helps assess how new processes, practices, and policies work in different universities as well as contributes to the social science knowledge base, especially regarding gender inequality and organizational innovations (Nelson and Zippel, 2021; Rosser et al., 2019; Zippel and Ferree, 2019). Over time, ADVANCE has expanded its focus to address systemic barriers due to other identities that intersect with gender to affect people's access and opportunities in academic STEM fields, including race/ethnicity, sexuality, class, and disabilities (Lee et al., 2022). An intersectional approach to understanding individuals and wider social contexts was incorporated into ADVANCE programming in 2016 (Morimoto, 2022).

With more than two decades of funding, NSF ADVANCE awardees have developed and implemented multiple strategies to address gender and intersectional equities. They include:

- Enhancing institutional structures through reviewing and revising policies for recruitment, hiring, promotion, and tenure as well as increasing the transparency and consistent implementation of these policies
- Providing work–life support through collecting and analyzing data regarding work–life issues, developing and implementing career policies accommodating various needs, and training administrators and faculty on these policies and work–life programs
- Improving equitable career support through establishing formal mentoring for faculty members and ensuring equitable allocation of resources by developing unbiased mechanisms to assign, track, and report teaching, service, and research
- Empowering faculty, department heads, and administrators through training on reducing implicit bias, creating research-based tools to reduce the impact of implicit

biases in decision making, and creating processes to track and evaluate the efficacy of the policies to achieve gender equity (NSF, 2021a; Stewart and Valian, 2018).

ADVANCE's impact remains after the funding period as it is estimated that two-thirds of funded institutions continue to support and promote activities developed by the ADVANCE team (DeAro et al., 2019). Some early ADVANCE organizations have institutionalized policies developed during the funding period a decade or more later and continue to introduce new measures and policies (Zippel and Ferree, 2019), and other organizations adapt models developed by or become allies of ADVANCE institutions. ADVANCE's impact is also found overseas. For instance, the INTEGER (Institutional Transformation for Effecting Gender Equality in Research) funding program in Europe was modeled after ADVANCE (DeAro et al., 2019).

NSF INCLUDES

To pursue diversity, equity, and inclusion among engineering students, NSF started the INCLUDES (Inclusion Across the Nation of Communities of Learners of Underrepresented Discoverers in Engineering and Science) program in 2017, one of NSF's 10 Big Ideas. INCLUDES and ADVANCE are both listed in NSF's 2022–2026 Strategic Plan, with the emphasis that students' demographic backgrounds should not limit their choices and chances of earning a STEM degree (NSF, 2022a). The five components of INCLUDES are: (1) shared vision, (2) partnership, (3) goals and metrics, (4) leadership and communication, and (5) expansion, sustainability, and scale. INCLUDES was renamed in August 2022 to the Eddie Bernice Johnson INCLUDES Initiative to honor this trailblazing U.S. Congresswoman, notable for her advocacy in STEM education and environmental issues (NSF INCLUDES National Network, 2022; Seeley, 2021).

An influential INCLUDES alliance is the Computing Alliance of Hispanic-Serving Institutions (CAHSI), which focuses on building and maintaining a community to enhance recruitment, retention, and advancement of Hispanics in computing (CAHSI, n.d.). CASHI started in 2006 with seven Hispanic-Serving Institutions (HSIs) funded by NSF's Broadening Participation in Computing. In 2016, with funding from the NSF INCLUDES Design and Development Launch Pilot program, CAHSI switched to a collective impact model (i.e., the idea that a network of committed institutions can achieve more than individual institutions). In 2018, CAHSI was selected by NSF to be one of the first five national INCLUDES Alliances, and now includes over 60 HSIs, industry partners, and other groups (Villa et al., 2019; Villa et al., 2020), and despite the COVID-19 pandemic, the alliance continued expand (Hug et al., 2021).

HSIs play an essential role in mentoring, networking, and preparing students for academic success and transition into the computing workforce (Gates, 2017). CAHSI supports student success by incorporating values grounded in the Hispanic community, including mutual support, respect for community members, and familial ties. Through programs such as Peer-Led Team Learning, CAHSI increases students' reported confidence in their ability to succeed in computing. The Fellow-Net program implemented at several CAHSI institutions provides opportunities for graduate students to work with mentors to enhance their grant applications (Núñez et al., 2021; Thiry, 2017).

CAHSI consistently exceeded national bachelor's degree graduation rates in computer engineering from 2002 to 2017. In 2006, the national graduation rate dropped to 64 percent of its 2002 level, while CAHSI institutions maintained a rate of 94 percent. By 2017, CAHSI institutions were graduating students at 148 percent of the 2002 rate, compared to 84 percent at other institutions. (Villa et al., 2019). CAHSI students have also reported a stronger sense of belonging than Hispanic computer science students at other institutions (Thiry, 2017).

INCLUDES focuses on community colleges as well, as they play an important role in transferring underrepresented minority students into 4-year colleges and engineering programs (Villa, 2017). Through the Aspire Alliance Regional Collaborative, Aspire fellows—graduate students from underrepresented groups in 4-year universities—are mentored to learn about effective and inclusive teaching skills as well as the culture of community colleges, thus encouraging fellows to consider rewarding career opportunities in community colleges (Flores et al., 2022).

NSF Broadening Participation in Engineering (BPE) Program

Since 2015, the NSF has funded projects under the Broadening Participation in Engineering (BPE) program, a program of the Division of Engineering Education and Centers in the Directorate for Engineering. To date, the NSF has invested nearly $175 million into the program, spanning 110 projects (NSF, n.d.-c).

The BPE program aims to foster a more inclusive engineering environment by funding projects that increase the participation of underrepresented groups. According to the description on the NSF website, "The BPE Program seeks to support not only research in the science of broadening participation and equity in engineering, but also collaborative endeavors which foster the professional development of a diverse and well-prepared engineering workforce as well as innovative, if not revolutionary, approaches to building capacity through inclusivity and equity within the engineering academic experience" (NSF, 2021b). To facilitate this goal, BPE supports projects in four tracks:

1. Planning and Conference Grants: Conference Grants engage communities and facilitate collaborations for future Planning Grants, while Planning Grants facilitate the development of collaborative BPE projects.
2. Research in Broadening Participation in Engineering: BPE-supported research provides evidence for engineering educators, administrators, employers, and policymakers to implement effective programs that broaden participation in engineering. This research identifies systemic barriers for underserved communities, develops methods to enhance access and retention, and aims to transform the culture towards diversity, equity, and inclusion across K-12 to professional levels
3. Inclusive Mentoring Hubs (IMHubs): This track supports proposals across engineering disciplines to create all-access, open-platform IMHubs that connect underrepresented racial and ethnic groups in STEM with mentoring and professional development opportunities. These IMHubs will cater to diverse communities, including students, educators, and professionals, aiming to establish sustainable networks over five years.
4. Centers for Equity in Engineering (CEE): CEE's aim to recruit and retain diverse students through systemic cultural, organizational, and pedagogical changes. Phase I of the CEE initiative focuses on establishing infrastructure and deploying inclusive practices within engineering colleges, while Phase II expands and sustains these efforts through partnerships with other institutions (NSF, 2021c).

Researchers and Educators Whose Work Has Been Supported by NSF

Gary S. May, Ph.D. is the Chancellor of the University of California, Davis. He was previously the Dean of the College of Engineering at the Georgia Institute of Technology. He is a member of the National Advisory Board of the National Society of Black Engineers and an elected member of the National Academy of Engineering. NSF has been supporting Dr. May's work since graduate school. He has published over 250 articles and technical presentations in semiconductor processing and the computer-aided manufacturing of integrated circuits (NASEM 2023). Dr. May also received the Presidential Award for Excellence in Science, Mathematics, and Engineering Mentoring in 2015 (White House, 2015).

Dr. May has had impact beyond research. With NSF support, he founded several programs aimed at broadening participation of historically underrepresented groups in STEM. Among them, the Summer Undergraduate Research in Engineering/Science (SURE) program at Georgia Tech (with $3 million in NSF funding) was designed to attract talented minority students into graduate school (SURE, 2024). As one of the longest running summer research programs in the country, SURE is tremendously successful in that about 75 percent of the students from the SURE program have enrolled in graduate school in engineering or science since the program inception in 1992 (Conrad et al., 2015; SURE, 2024).

The success of the SURE program set the foundation for another, much larger NSF grant for the Facilitating Academic Careers in Engineering and Science program (FACES[66]). FACES was designed to increase the number of African American students receiving doctoral degrees from Georgia Tech and then launching their academic STEM careers and becoming role models. FACES helped produce over 400 minority doctorate recipients in STEM, which surpassed all other universities in the whole country over the time period (1998–2013) (NASEM, 2023).

Karan Watson, Ph.D., is the Regents Professor in the Department of Electrical and Computer Engineering at Texas A&M University and has previously served in administrative roles there such as provost and executive vice president, vice provost, associate Dean for graduate studies in the College of Engineering, President of ABET (Accreditation Board for Engineering and Technology), President of the Education Society of IEEE, and more. Professor Watson is a fellow of IEEE, ASEE, and ABET. Her awards and recognitions include the U.S. President's Award for Mentoring Minorities and Women in Science and Technology, AAAS mentoring award, IEEE International Undergraduate Teaching Award, College of Engineering Crawford Teaching Award, and two Distinguished Achievement Awards from the Texas A&M University Association of Former Students in student relations (1992) and in administration (2010) (Texas A&M University, n.d.). Most recently, she was awarded the 2021 ASEE Lifetime Achievement Award in Engineering Education for "her pioneering leadership and sustained contributions to education in the fields of engineering and engineering technology" (Meyers, 2021). Watson was the first woman to graduate with a Ph.D. in engineering from Texas Tech in 1982.

Since the start of her career, Watson has been a champion for underrepresented populations in engineering. In 1990, she was a PI on a 5-year NSF grant that brought in over $1.2 million dollars to Texas A&M for "Graduate Engineering Education for Women, Minorities and/or Persons with Disabilities".[67] Watson spearheaded the Texas Alliance for Minority Participation from 1991 to 2007 in which Texas A&M led a coalition of 9 Texas universities, 31

---

[66] Award #0450303.
[67] Award #9017249.

community colleges, 67 industrial partners, and 12 national laboratories "dedicated to improving the quality of the undergraduate and graduate education of minorities in engineering, mathematics and the sciences through a program of enhanced preparation, recruitment, transfer and retention" (NSF, n.d.-f), bringing in nearly $11.4 million in NSF funding. Watson led numerous other projects such as "Changing Faculty through Learning Communities," which focused on evolving the faculty culture through faculty training in strategic disciplines to create inviting and inclusive learning environments. From 2003 to 2013, Watson was PI for the Texas A&M University System Louis Stokes Alliance for Minority Participation (LSAMP), which focuses on increasing the number of STEM bachelor's and graduate degrees awarded to underrepresented populations (NSF, n.d.-d). As PI, Watson secured nearly $13.5 million in funding from NSF for LSAMP, not to mention her contributions as co-PI from 2013 onward. At the time of the writing of this report, Texas A&M's LSAMP program had supported 36,000 students through their various programs including 68 Ph.D.s (TAMUS LSAMP, n.d.).

Sarah EchoHawk, a citizen of the Pawnee Nation of Oklahoma, has devoted more than 20 years to advancing Indigenous communities, notably in the realm of STEM education. Serving as the chief executive officer of the American Indian Science and Engineering Society (AISES) since 2013, she has been at the forefront of the organization's efforts to enhance Indigenous representation in STEM fields (AISES, 2022). Established in 1977, AISES currently has over 6,000 members and supports a vast network of 230 affiliated pre-college schools, 196 college and university chapters, 3 tribal chapters, and 18 professional chapters spanning the United States and Canada (AISES, 2016; NASEM, 2023). AISES offers numerous benefits to its members, including over $13 million in academic scholarships, internships, professional development resources, and national and regional conferences.

Additionally, NSF has supported AISES through INCLUDES grants and in conducting research to identify the factors influencing the persistence and success of Indigenous scholars and professionals in STEM (Page-Reeves, 2017). This support includes initiatives such as the 50K Coalition and Engineering Plus awards, which aim to boost the annual number of engineering bachelor's degrees awarded to women and underrepresented minorities in the United States from 30,000 to 50,000 by 2025, constituting a 66 percent increase. Further programs, such as the Innovation Technology Experiences for Students and Teachers (ITEST) initiative, target engagement of Indigenous girls in computer science, while efforts such as "Lighting the Pathway to Faculty Careers for Natives in STEM"[68] examine the experiences of Native learners and professionals to enhance their representation in STEM faculty roles nationwide. Prior to EchoHawk's role at AISES, she held positions at First Nations Development Institute and the American Indian College Fund, demonstrating extensive experience in nonprofit management and advocacy for Indigenous causes. Her commitment to education is exemplified by her years as an adjunct professor of Native American Studies, fostering academic growth and cultural understanding. Furthermore, her advocacy extends beyond academia through involvement in various boards and committees, such as the American Indian Policy Institute and the Last Mile Education Fund, emphasizing the importance of inclusive research and collaboration with Indigenous communities to ensure their meaningful participation in shaping their educational and professional journeys.

NSF funding has thus empowered colleges, universities, and other organizations committed to achieving gender and intersectional equities as well as individual researchers and their teams to achieve their goals of broadening participation in engineering and science. A

---

[68] Award #1935888.

diverse engineering workforce benefits not only their fields but also society as a whole, as is showcased in the tremendous social impacts they have been making. Moving forward, NSF's role in nurturing talent and fostering a culture of discovery and innovation will undoubtedly continue to be a cornerstone of progress and engineering excellence. The legacy of NSF's support is a testament to the power of investing in the future, ensuring that the field of engineering remains vibrant, dynamic, and at the forefront of shaping our world for the better.

MATERIALS SCIENCE AND ENGINEERING

Materials science and engineering have long been instrumental in shaping the modern world and underpinning technological and societal advancements. The scientists and engineers in these disciplines seek to comprehend and manipulate the fundamental properties of physical materials, with the ultimate goal of enhancing existing materials and creating novel materials for innovative applications. The impact of materials science and engineering innovations has been transformative across a multitude of sectors, including energy, transportation, health care, electronics, construction, and manufacturing. Moreover, the concerted federal effort to advance materials science and engineering in the latter half of the 20th century in the United States not only supported these disciplines but also paralleled the broader advancement of science and engineering.

The launch of Sputnik by the Soviet Union in 1957 ignited the "space race" and triggered a domino effect of events that led to the rapid expansion of materials science and engineering research in the United States. These events included a 1958 report by the President's Science Advisory Committee entitled "Strengthening American Science," which advocated that "a special institute should be created to work exclusively on new metals and materials, if we are to obtain the strength and heat resistance demanded by our unfolding technologies—both military and nonmilitary" (President's Science Advisory Committee, 1958, p. 7). Subsequently, the Federal Council for Science and Technology was established, appointing a Coordinating Committee on Materials Research and Development (CCMRD). The CCMRD established the Interdisciplinary Laboratories (IDL) program for materials science and engineering through the Advanced Research Projects Agency (ARPA[69]), which was launched in 1960 (NRC, 1975). At this time, materials science and engineering did not exist as a discipline. It was formerly encompassed in metallurgy or ceramics engineering programs. The IDL program ran for more than a decade before it moved to the National Science Foundation (NSF) and was reestablished as the Materials Research Laboratories (MRL) program. The IDL and MRL programs thus helped launch this completely new, interdisciplinary field (Baker, 1987).

From 1972 to 1996, the NSF administered the MRL program, introducing an innovative funding model at the agency (NRC, 2007). Traditionally, NSF had awarded individual grants for research confined to specific disciplines. However, it was recognized that comprehensive materials science and engineering research necessitated interdisciplinary collaboration from experts in fields such as physics, chemistry, and mathematics. Accordingly, the MRL program issued block-type grants for multidisciplinary research and shared facilities, as was true for the IDL program under ARPA. William Baker (1915–2005), former president and chairman of Bell Telephone Laboratories, wrote a 1987 historical perspective for the National Academy of Sciences on "Advances in Materials Research and Development" (Baker, 1987). In this piece, Baker underscored the importance of the MRL program's model as a frequently overlooked

---

[69] The agency changed its name to the Defense Advanced Research Projects Agency (DARPA) in 1972.

precursor to the NSF's highly successful Engineering Research Center (ERC) program, established in 1985.

From 1994 to 1996, the MRL program transitioned into the NSF Materials Research Science and Engineering Centers (MRSEC) program. The MRSEC program is ongoing, and, as of the writing of this report, there are 28 active MRSECs. The most recent MRSEC competition in 2023 awarded 9 grants totaling $162 million (NSF, 2023e). Because of the interdisciplinary nature of materials science and engineering, NSF support for the field is also evident in myriad other programs and initiatives, including the ERCs. This support has helped to spur innovations such as:

- **Development of sustainable biotextiles:** The mainstream fashion and textile industry has long been criticized for its unsustainable cradle-to-grave model, which results in significant volumes of waste and contamination from petroleum-based textiles and synthetic dyes. In 2017, researchers at Columbia University's MRSEC achieved first place in the Sustainable Planet category of the National Geographic Chasing Genius challenge with their startup AlgiKnit (NSF MRSEC, 2017) This company makes kelp-based biodegradable knitwear, apparel, and footwear with natural pigment and a closed-loop lifecycle. In 2020, researchers at this same MRSEC unveiled "compostable bioleather" inspired by pre-industrial and indigenous science, which relies on the microbial biosynthesis of nanocellulose (Kelso, 2022). Another startup from this MRSEC, known as Werewool, focuses on developing textile fibers with inherent color and performance properties reliant on biomimicry protein structures instead of synthetic dyes and fibers (Werewool, 2020). Biotextile alternatives to fast fashion hold promise for moving the needle closer towards global sustainability goals and a circular economy.

- **Biocompatible medical implants and materials:** The NSF ERC for Revolutionizing Metallic Biomaterials (ERC-RMB), led by North Carolina A&T State University in collaboration with the University of Pittsburgh and the University of Cincinnati[70], has pioneered the development of biomaterials and "smart" implants for various medical interventions, such as craniofacial, dental, orthopedic, cardiovascular, thoracic, and neural procedures. One of the center's major achievements has been the development of biodegradable metals, particularly magnesium-based alloys, with the aim of creating implants that can adapt to the human body and eventually dissolve when they are no longer needed. This innovation has the potential to reduce the need for multiple invasive surgeries and lower health care costs. The ERC-RMB has also contributed to advances in materials processing and characterization, modeling, biocompatibility testing, and the use of data mining, machine learning, and artificial intelligence in biomaterials manufacturing. The center has promoted entrepreneurship and diversity in engineering education, prioritizing a "culture of inclusion," and it was the first ERC to be based at a Historically Black College or University (NSF, 2008).

- **Recycling of lithium ion (Li-ion) batteries for electric vehicles (EVs):** In 2022, President Biden invoked the Defense Production Act to spur the domestic recycling of EV batteries, given that "the U.S. and its allies currently do not produce enough of

---

[70] Award #0812348.

the critical minerals and battery materials needed to power clean energy technologies" (White House, 2022b). However, rewinding to 2012, Worcester Polytechnic Institute professors Yan Wang and Eric Gratz faced challenges securing funding for fundamental research recycling of Li-ion battery materials, primarily due to the nascent state of the technology and limited expectations concerning its necessity (NSF, 2022b). Fortunately, the team was able to secure initial research investment from the Center for Resource Recovery and Recycling, an NSF Industry-University Cooperative Research Center (IUCRC). With additional backing from the NSF Innovation Corps (I-Corps) program and seed funding from NSF's Small Business Innovation Research (SBIR) program, the team launched the company Ascend Elements, which grew and inaugurated North America's largest electric vehicle battery recycling facility in May 2023 (Ascend Elements, 2023).

When looking at important figures in the history of materials science and engineering, particularly nanoscience, it would be difficult to overstate the importance of the work of Mildred (Millie) Dresselhaus (1930–2017).[71] In 2014, President Obama awarded Dresselhaus the Presidential Medal of Freedom, stating that "her influence is all around us, in the cars we drive, the energy we generate, and the electronic devices that power our lives" (Hunter College, 2014). Dresselhaus enjoyed a long and diverse career as an Institute Professor—the highest academic title—at the Massachusetts Institute of Technology (MIT), where her research encompassed the study of carbon and semi-metals. Her career began with an NSF-sponsored fellowship in 1958, dedicated to the exploration of superconducting materials before her tenure at MIT. Upon joining MIT in 1960, she shifted her focus to the fundamental properties of carbon atoms—a field that received limited attention at the time. By delving into the electronic structure of graphite and carbon as a whole, Dresselhaus laid the foundation for a wealth of new developments in science and engineering. Her work contributed to the discovery of carbon nanotubes, which are essential in developing stronger and lighter materials for aerospace and construction; buckyballs, which have potential applications in drug delivery systems and materials science; and the advancement of quantum computing, paving the way for ultra-fast and secure data processing and revolutionizing fields such as cryptography, artificial intelligence, and complex system simulations. Her work on superlattice structures enabled the technologies leading to lithium-ion batteries[72] that are used in electric vehicles and renewable energy storage (NIHF, 2014). Beyond her groundbreaking research, Dresselhaus was a cherished educator and a strong advocate for increased participation and mentorship of women in STEM. Her research endeavors received support from 20 different NSF grants.

Within the extensive landscape of researchers whose careers have been shaped by the foundational work of Mildred Dresselhaus, one notable figure is Dr. Baratunde Cola. As a professor at the Georgia Institute of Technology, Cola's research portfolio encompasses critical domains such as heat transfer, combustion and energy systems as well as micro and nano engineering (Georgia Tech, n.d.). One highlight of Cola's research was the use of carbon nanotubes for heat dissipation, culminating in the scalable production of organic and organic–inorganic hybrid nanostructures with diverse technological applications. These cutting-edge

---

[71] Maia Weinstock's presentation on the life and work of Mildred Dresselhaus is summarized in the proceedings of the National Academy of Engineering's 2022 symposium on Extraordinary Engineering Impacts on Society (NASEM, 2023).

[72] U.S. Patent No. 7,465,871.

technologies include materials that manage heat transfer, like devices that convert heat into electricity, antennas that capture and convert light energy, strong materials made from tiny carbon tubes, and adjustable materials that precisely control heat flow. It is important to recognize that these research pursuits in the field of heat transfer address universal challenges, often underappreciated particularly in the context of clean energy and the efficiency of everyday technologies that Americans rely upon. Cola's journey to prominence began with his achievements as an athlete, where he excelled as a starting fullback during his undergraduate years at Vanderbilt University (Vanderbilt University, 2017a). Subsequently, he firmly established himself as one of the nation's foremost young engineers, exemplified by his receipt of the 2012 Presidential Early Career Award for Scientists and Engineers presented by President Obama (White House, 2012), and the 2017 NSF Alan T. Waterman Award—the highest honor bestowed upon early-career scientists and engineers in the United States (Vanderbilt University, 2017b). Cola directs the Georgia Tech Nanoengineered Systems and Transport Lab, which has led to the creation of his startup venture, Carbice (Armstrong, 2022). This company specializes in the production of carbon nanotubes for a wide spectrum of electronics cooling applications.

A third example of a materials science and engineering researcher using nanotechnology is Paula Hammond, Institute Professor, Vice Provost for Faculty, and former Department Head of Chemical Engineering at MIT. As a chemical engineer, she embraces the opportunity to "manipulate matter in new and exciting ways, to be able to build something truly incredible" (MIT, n.d.-a). In the early stages of her career, Hammond secured funding from NSF for her work involving the use of oppositely charged polyelectrolytes to construct thin films. These films, constructed one nanolayer at a time, "encapsulate the drug like shrink wrap" (Shen, 2024). Subsequently, her research endeavors extended to the study of how these materials assemble in solution and the use of synthetic polypeptides as carriers for pharmaceutical agents (NASEM, 2023). Later, Hammond, in collaboration with her colleagues at MIT, harnessed FDA-approved nanoparticles to encapsulate and deliver chemotherapy drugs with precision to specific target cells, addressing the formidable challenges posed by lung, breast, and ovarian cancer. In her presentation at the NAE Symposium on Extraordinary Engineering Impacts on Society in 2022, Hammond shed light on the unique challenges associated with ovarian cancer, emphasizing its often late stage detection. Through a cataloging of different kinds of nanoparticles with varying outer-layer charges, Hammond and her team identified three compositions with a remarkable affinity for ovarian cancer cells. This breakthrough not only holds promise for early detection but could also train the immune system to recognize ovarian tumors before they begin to grow. A substantial portion of Hammond's fundamental research in polymer science has received support from NSF, and her distinction as one of the very few individuals to be elected to all three of the National Academies—The National Academy of Sciences, the National Academy of Engineering, and the National Academy of Medicine—underscores the impact of her contributions to many diverse fields.

In essence, the story of materials science and engineering is one of continuous innovation and interdisciplinary collaboration, driven by a shared commitment to advancing science and technology for the betterment of society. Looking to the future, these disciplines will continue to play a crucial role in addressing some of society's most pressing challenges and driving progress in a wide range of fields.

NSF CENTERS (ENGINEERING RESEARCH CENTERS AND OTHERS)

President John F. Kennedy's exhortation for the country to become an international leader in space exploration and to go to the moon, not because it was easy, but because it was hard, still resonates today as an enduring source of motivation and inspiration. These words underline the importance of tackling ambitious goals that push the boundaries of human achievement. They are clearly relevant to the NSF's centers initiatives and to the engineering profession as they embody a shared vision, determination to respond to unprecedented challenges, unwavering focus, strategic allocation of resources, acceptance of high risks and failure, the power of nontraditional teams, and recognition of hidden talents. The United States' momentous journey to the moon within 7 years after Kennedy's call to action exemplifies the power of these principles, which continue to shape the DNA of NSF centers and drive transformative achievements. This section underscores the vital role of centers within the NSF portfolio and their continuing extraordinary contributions to engineering research and education.

The adoption of a centers model at NSF faced skepticism in the 1980s. The agency—initially established to support basic research through grants to individual principal investigators or small teams—was unaccustomed to an approach that demanded higher funding levels, cross-disciplinary and multiorganization collaboration, and administrative oversight extending up to a decade. Nevertheless, the model has withstood the test of time at NSF, and subsequently it has been strategically adopted at other federal agencies. As NSF's inaugural holistic center concept, the Engineering Research Center (ERC) program has effectively fulfilled its mission to address U.S. competitiveness and prepare students for engineering practice in industry. A 2010 study estimated that $50–75 billion in downstream economic value had been generated from technologies developed by ERCs in the first 25 years of the program (Preston and Lewis, 2020).

The embrace of the centers model by NSF is a narrative woven together by geopolitical events and the evolution of the engineering profession. During the Cold War, pressures stemming from global competitiveness, the quality of public education, and the need for scientific and engineering accomplishments led to a series of federal actions that elevated the prominence of engineering at NSF. Over time, it became evident that engineering achievements, rather than fundamental scientific discoveries, formed a primary basis for the agency's case to Congress for augmented budgets. Key events in these early years included:

- The 1968 Daddario–Kennedy amendment to the National Science Foundation Act of 1950 (Public Law 90-407) expanded the NSF charter to include applied research in addition to basic research.
- The Research Applied to National Needs (RANN) program (1971–1978), which stemmed from the Daddario-Kennedy amendment, focused on economic and socially relevant research to address domestic challenges such as pollution, energy, and global competitiveness (NSF, n.d.-g, 1994; Preston and Lewis, 2020).
- The founding of RANN's Industry–University Cooperative Research Centers (IUCRC) program in 1973 fostered industry–academia collaboration (NSF, n.d.-a).
- The early 1980s proposal to create a separate National Engineering Foundation (Bozeman and Boardman, 2004) in response to tensions between basic science and applied engineering which kept funding for engineering low (NSF, 1994).

Although the National Engineering Foundation did not come to pass, NSF established the Directorate for Engineering (ENG) in March 1981, significantly increasing the profile of engineering at NSF. ENG soon conducted an assessment of engineering education and research, culminating in a recommendation to establish on-campus "engineering centers" to support emerging research fields, establish industry partnerships, and facilitate cross-disciplinary research. The National Academy of Engineering (NAE) helped design the ERC framework (NAE, 1983) at the request of NSF, with direct engagement by universities, industry, and government, and in 1985—under the leadership of NSF Director Erich Bloch—the ERC program was officially launched. This marked a seminal development, occurring just four years after the inception of the engineering directorate.

The ERC program went on to become a hallmark of the directorate and a gold standard for NSF. ERCs have been instrumental in producing breakthroughs that address national challenges, launch entirely new industries and fields of study, and create new and foundational opportunities. The program also introduced a tool to help organize research strategies by breaking down ambitious 10-year visions into manageable components: the three-plane diagram (Figure 4-2). The diagram consists of "systems," "enabling technologies," and "fundamental knowledge" planes and serves as a framework guiding ERCs and other centers in identifying barriers to achieving 10-year visions and developing research plans to overcome them. Although other center programs at NSF do not formally require the three-plane chart, they have adopted many elements of this tool.

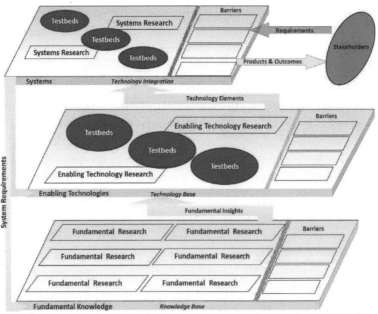

FIGURE 4-2 The National Science Foundation's "Three-Plane Diagram," the basis for strategic planning for engineering research centers (NSF, 2020a; Figure 3-2).[73]

Because of persistent concerns about global competitiveness and education, ENG and other directorates at NSF continued to introduce networks and other center-like programs that advanced engineering, education, and science throughout the United States. Examples of these are noted below:

---

[73] For more on the three-plane diagram, see (NASEM, 2023 pp. 53-55).

- The Science & Technology Centers (STCs) program was announced by NSF in 1987, launched in 1989, and continues to this day. STCs support innovative, large-scale research projects which are intended to foster breakthroughs, foster collaboration with industry, generate transformative technologies, and train scientists. STCs focus on creating new scientific paradigms through interdisciplinary partnerships and contribute to global leadership in research and education across diverse science and engineering areas. The program has supported 62 centers to date (NSF, 2023f).

- In 1994, NSF launched the National Nanofabrication User Network (NNUN), a national network of user research facilities to connect expensive facilities, foster synergy, and reduce duplication. In response to President Clinton's 2000 National Nanotechnology Initiative, a series of NSF programs were launched including the Nanoscale Science & Engineering Centers (NSECs; established in 2001), and the Network for Computational Nanotechnology (2002). In addition, the NNUN concept was broadened and recompeted to establish the National Nanotechnology Infrastructure Network (2003). These programs collectively constitute a dynamic framework for pioneering nanoscale research, fostering computational nanotechnology advancements, and providing extensive infrastructure support to propel the field of nanotechnology into the future (NSF, 2003, 2006; Roco, 2011).

- 1994 also marked the inception of the Packing Research Center at Georgia Tech. This ERC facilitated research and development of "System-in-package" (SiP) packaging, a means of integrating multiple, diverse electronic components into a single module. SiP technology is used in smartphones and other devices where space is at a premium. A study estimated that the state of Georgia, which had invested $32.5M in the Center between 1994 and 2004, realized a yield of $192M in direct economic impacts and an additional $159 million in indirect and induced impacts (SRI International, 2008).

- The Earthquake Engineering Research Centers were funded in 1997 in response to Congress's Earthquake Hazards Reduction Act of 1977 (Public Law 95-124). The George E. Brown, Jr., Network for Earthquake Engineering Simulation was established in 2004. These centers and network are dedicated to advancing seismic resilience nationwide through innovative research and collaboration, including site remediation, structural control and simulation, high-performance materials, and decision support systems (NSF, 2000b), and their efforts have contributed to protecting property and saving lives.

- The National Artificial Intelligence Research Institutes (AI Institutes) program was established by NSF's Computer and Information Science and Engineering (CISE) directorate in 2020 after a 2019 Executive Order on AI (E.O. 13859[74]). A partnership among federal agencies and private sector organizations led by CISE supports fundamental research, education, and workforce development in this fast-growing and vital field. As of 2023, there were 25 NSF-funded AI Institutes in 40 states and the District of Columbia, with a combined $220 million NSF investment (NSF, 2023a). In October 2023, Executive Order 14110—Safe, Secure, and Trustworthy Development

---

[74] https://www.federalregister.gov/documents/2019/02/14/2019-02544/maintaining-american-leadership-in-artificial-intelligence.

and Use of Artificial Intelligence[75]—was issued, initiating a government-wide program aimed at steering responsible AI development and implementation through leadership from federal agencies, industry regulation, and collaboration with international partners. NSF will also have a major role in the implementation of directives contained in this order.

- The Partnerships for Research and Education in Materials (PREM) program, initiated in 2003 by the Division for Materials Research, supports Minority-Serving Institutions (MSIs) in partnership with Materials Research Science and Engineering Centers (MRSEC). This collaboration empowers an MSI to lead a major center program and engages the underrepresented communities in cutting-edge materials science and engineering research and education (Partnerships for Research and Education in Materials, n.d.).

- The Engineering Directorate's Division of Engineering Education and Centers[76] oversees education-based initiatives within the ERCs and the Industry-University Cooperative Research Centers, including those related to research and practical experience opportunities for students and teachers and research in engineering education. The chapter section titled "Engineering Education and Early Career Development" provides details on some of the division's signature programs: Broadening Participation in Engineering (BPE), Faculty Early Career Development Program (CAREER), and NSF INCLUDES (Inclusion across the Nation of Communities of Learners of Underrepresented Discoverers in Engineering and Science). One example of the agency's Research Experiences for Undergraduates (REU) program is described in the chapter section titled "Wind Energy".

While other federal agencies, such as the DoD with its Centers of Excellence at Minority-serving Institutions (DoD, 2023), have also adopted the centers model and often collaborate with the NSF, the NSF's broad support across a diverse range of scientific and engineering disciplines uniquely positions it to sustain interdisciplinary centers with engineering capabilities. This support spans many disciplines and extends over time. Figure 4-3 presents a timeline of significant events related to NSF centers through 2020.

---

[75] https://www.whitehouse.gov/briefing-room/presidential-actions/2023/10/30/executive-order-on-the-safe-secure-and-trustworthy-development-and-use-of-artificial-intelligence/.
[76] https://www.nsf.gov/eng/eec/about.jsp.

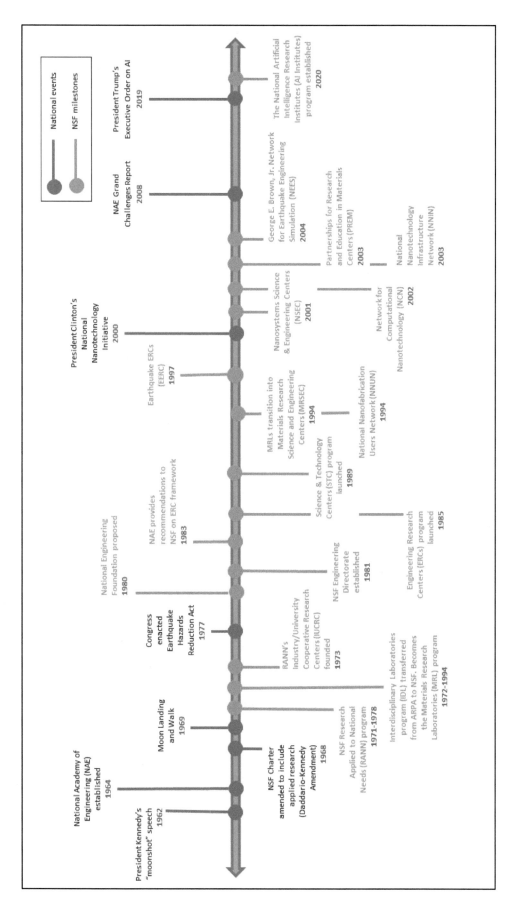

FIGURE 4-3 Timeline of events relevant to NSF centers through 2020.

It is worth noting that the first center-like program under NSF's purview was the Materials Research Laboratories (MRL) program, transferred to NSF from the Advanced Research Projects Agency (ARPA) in 1972 and formerly known as the Interdisciplinary Laboratories (IDL) program (Sproull, 1987). These shared-equipment facilities were intended to support interdisciplinary materials research and to train personnel on state-of-the-art equipment. While MRLs had different missions and operating structures than ERCs, observers have argued that they served as a model for future NSF programs (Baker, 1987; NRC, 2007). In fact, the MRL program was later converted into the MRSEC program in 1994 by integrating formal research studies that required use of those MRL facilities. The historical trajectory of the MRLs follows a recurring pattern where technologies or programs may originate in other government agencies before being transferred to NSF once they mature or due to policy considerations (see for instance, the story of the internet or the transfer of MOSIS from DARPA to NSF, both of which are discussed elsewhere in this chapter). Nevertheless, ERCs embodied the first holistic centers concept that originated in NSF (i.e., grand vision, strong cross-disciplinary research, transformative engineering education, and non-academic partnerships).

In 2022, the report's authoring committee conducted a symposium on "Extraordinary Engineering Impacts on Society" (NASEM, 2023). This symposium included a session on NSF centers featuring speakers who were current or former directors of center types mentioned above. The session's speakers identified many common characteristics and advantages of these center programs which enable large teams to:

- realize transformative visions and address complex engineered system challenges in collaboration with industry and other stakeholders;
- reconceptualize engineering education with cross-disciplinary and applied programs and opportunities for engineering practice through internships;
- promote interdisciplinary and inclusive cultures that broaden participation in engineering;
- demonstrate complex engineered systems integration at scale, including testbed development, system-level performance demonstration, and process optimization;
- establish enduring partnerships with industry and global stakeholders;
- cultivate innovation ecosystems;
- transform academic culture;
- create value and drive meaningful impact through new industry launches and regulatory/policy innovations; and
- develop leaders with the capacity to see the big picture and to design and execute on large-scale programs.

As of late 2023, NSF had provided support to 79 ERCs. These initiatives have yielded impressive results, including the publication of over 25,000 peer-reviewed journal articles and books; the acquisition of more than 800 patents, 1,300 licenses, and 2,500 invention disclosures; the establishment of over 240 spinoff companies; and the conferral of more than 14,400 bachelor's, master's, and doctoral degrees to ERC students (NSF, 2023d). The committee's research and responses to questionnaires circulated to NAE Members yielded a number of noteworthy examples of these, several of which are described below.

The Collaborative Adaptive Sensing of the Atmosphere (CASA) ERC. CASA was founded in 2003 and is headquartered at the University of Massachusetts, Amherst. The ERC's achievements include the development of low-cost collaborative adaptive radar networks. These radar systems have significantly improved weather forecasting, enabling more precise storm characterization and early warnings for severe weather conditions. CASA's radar network has played a crucial role in providing data during weather events, potentially saving lives. Moreover, its focus on student involvement has provided educational and research opportunities, fostering future expertise in the field of atmospheric sensing. After its founding, CASA later transitioned from tornadoes to severe storms and developed the CASA Dallas–Fort Worth Living Lab in partnership with North Texas Public Safety to conduct research on end-to-end severe weather warnings and human response (NSF, 2014; University of Massachusetts, 2023).

The ERC for Emerging Cardiovascular Technologies (ERC–ECT). This ERC, established at Duke University in 1997, focused on preventing sudden cardiac death through innovative research in implantable defibrillators and related technologies. The advances it made in electrode technology and biphasic waveforms have been adopted by industry leaders such as Medtronic and have improved device performance. These innovations have not only enhanced the effectiveness of implantable defibrillators but have also led to the development of portable defibrillators, benefiting individuals experiencing heart attacks in public places. Additionally, the ERC–ECT's research in three-dimensional ultrasound technology laid the foundation for its widespread adoption, with broad applications in the medical field (Lewis, 2010; NSF, 2015b).

The Data Storage Systems Center (DSSC) ERC. DSSC was established at Carnegie Mellon University (CMU) in 1990 and has had a significant impact on data storage technology. Under the leadership of Mark Kryder, the ERC built upon innovations in perpendicular magnetic recording (PMR) technology, which surpassed the limitations of longitudinal recording and substantially increased data storage density. Dr. Kryder transitioned from CMU to the data storage technology company Seagate—maintaining a close relationship with the DSSC—where PMR was implemented in hard disk drives (HDDs) and successfully commercialized. By 2005, PMR became the industry standard. This breakthrough helped sustain the growth of HDDs, which play a critical role in the cloud computing industry due to their cost-effectiveness. Moreover, the DSSC's research in heat-assisted magnetic recording has further extended technological advances in data storage, ensuring the continued relevance of HDDs (CMU, 2016; NSF, 2000a).

The Center for Subsurface Sensing and Imaging Systems (CenSSIS) ERC. CenSSIS is an ERC that took the three-plane chart to heart. CenSSIS was in the ERC Class of 2000. It was a multi-institution center led by Northeastern University in partnership with Boston University, Rensselaer Polytechnic Institute, the University of Puerto Rico at Mayagüez, and several strategic affiliates. The CenSSIS team developed generalized physics-based signal processing models to probe the subsurface at different wavelength regimes, from sub-cellular (100 nm–100 μm) to sub-sea (10 cm–1 km). CenSSIS was able to use this research to win a Department of Homeland Security (DHS) center of excellence award to detect and mitigate explosives, framing the center goals with the three-plane chart. After 10 years, the outcomes of this DHS center were applied to yet another DHS center to neutralize threats in soft targets–crowded places environments (e.g., sports venues, schools, etc.). CenSSIS leadership then developed a large-scale strategy for an NSF INCLUDES Alliance: Engineering PLUS. The idea was to use this tool to take a systems approach to unraveling the challenges of broadening participation in engineering and to create a national alliance that could collectively achieve the long-term, sustainable vision of inclusion.

The Biomimetic Microelectronic Systems (BMES) ERC[77] was led by Dr. Mark Humayun, a distinguished professor at the University of Southern California, in collaboration with the California Institute of Technology and the University of California, Santa Cruz. The BMES ERC was intended to develop biomimetic microelectronic systems to address critical challenges, including restoring lost sensory functions. The center's three testbeds focused on creating interfaces with the human nervous system for sight restoration, developing systems for cognitive function restoration, and designing a cellular testbed for light-sensitive neurons. Facing obstacles such as preserving microelectronics in the corrosive environment of the human body, the ERC successfully developed an Food and Drug Administration–approved implanted device to treat advanced retinitis pigmentosa, marking a breakthrough in vision restoration. The BMES ERC showcased a significant university–industry collaboration involving multiple disciplines which generated numerous patents, publications, and startup companies. Post-NSF funding, it transitioned into the Ginsburg Institute for Biomedical Therapeutics, continuing its beneficial work in converting technologies into therapies and training the workforce needed for their engineering and commercialization. The institute's ongoing projects, such as an implanted eye camera, demonstrate a commitment to advancing neural function restoration on a potentially transformative scale.

As captured in the previous examples, centers and networks are natural ecosystems for developing engineering leaders with the capacity for large-scale initiatives and grand challenges. Other notable leaders of NSF centers include Dr. Veena Misra, a Distinguished Professor of Electrical and Computer Engineering at North Carolina State University, who currently serves as the Director of the NSF NSEC on Advanced Self-Powered Systems of Integrated Sensors and Technologies (ASSIST), established in 2012. The ASSIST center has had a material impact on global health by using nanotechnology to create self-powered, wearable health monitoring systems, enabling long-term health management and improved quality of life outcomes. Their work advances environmental health research, clinical trials, and public health, while also promoting STEM education and diversity in engineering careers.

Dr. Misra's journey, closely intertwined with the ERC program, began as a graduate student conducting her Ph.D. research at the Gen-1 ERC for Advanced Electronic Materials Processing at NCSU. Following a stint in industry at the Motorola Advanced Products Research and Development Laboratories, she joined NCSU's faculty in 1998. She continued her involvement in ERCs, taking on a leadership role in the Gen-2 ERC for Future Renewable Electric Energy Delivery and Management Systems Center, launched in 2008. The center has advanced the development and commercialization of advanced energy distribution technologies with innovations in energy routing, solid state transformers, and control algorithms that revolutionize how electric utilities interface with customers on the grid, leading to a more efficient and flexible energy delivery system, numerous patents, and the establishment of spin-off companies. Her funding includes a 2001 NSF CAREER Award and a 2001 Presidential Early Career Award for Scientists and Engineers. Dr. Misra's contributions span research, teaching, and her impact on diversity, equity, and inclusion, earning her election as an Institute of Electrical and Electronics Engineers Fellow, among other honors.

---

[77] Further information on the Biomimetic Microelectronic Systems ERC is contained in the chapter section titled "Biomedical/Rehabilitative Engineering".

Dr. Karen Lozano is the Julia Beecherl Endowed Professor of Mechanical Engineering at the University of Texas–Rio Grande Valley (UTRGV) and has used her NSF center to make impacts on broadening participation in engineering. She has been the Principal Investigator and Director of a PREM center in partnership with the University of Minnesota MRSEC that specializes in carbon nanofiber-reinforced thermoplastic composites. More specifically, these centers encompass three Interdisciplinary Research Groups aiming to (1) advance comprehension of charge transport within solid-state materials, which plays a pivotal role in technologies such as plastic electronics and magnetic storage devices; (2) create non-toxic nanocrystals from readily available elements, with the aim of crafting thin films suitable for applications in solar energy and low-energy lighting; and (3) develop innovative methods for assembling polymeric materials with exceptional property combinations, facilitating a wide range of applications spanning water treatment, fuel cell membranes, gene therapy, and integrated circuit manufacturing.

Dr. Lozano's university, UTRGV, is a Minority-Serving Institution with approximately 90 percent Hispanic, primarily first-generation students. The multi-year duration of the PREM center has enabled Lozano to create a culture of achievement for her students, strategically pursuing projects that allow each student to make realistic contributions. Roughly 85 percent of her more than 200 peer-reviewed journal articles have been co-authored by undergraduate students. She has mentored over 500 undergraduate students with a 100 percent graduation rate in engineering. Many of these students have continued their studies in graduate school. Dr. Lozano has more than 20 patents and patent applications, and the students participate in these entrepreneurial activities with her as well. Notably, despite her university's focus on undergraduate education rather than research, she received a 2000 NSF CAREER Award, two Major Research Instrumentation Program grants, and other significant NSF research grants. Dr. Lozano has earned multiple awards and honors, including the 2019 Presidential Award for Excellence in Science, Mathematics, and Engineering Mentoring and the distinction of being elected to the 2020 class of the National Academy of Inventors and the 2023 class of the National Academy of Engineering (NAE), making her the first NAE member to achieve that status while building her entire career at a non-R1 institution[78].

In conclusion, the NSF centers and those spearheading them epitomize a shared vision, the resolve to tackle unprecedented challenges, unwavering focus, and the power of diverse teams. These principles have guided NSF centers in their transformative endeavors, producing extraordinary contributions that affect society, address national challenges, launch new industries, and create extensive opportunities for students and researchers. From the early days of the innovative ERC program to the diverse center models and alliances embraced across all NSF directorates today, the centers model has endured and thrived. It is a testament to the continued relevance of engineering centers in shaping the landscape of science and engineering. Just as President Kennedy's moonshot speech was a defining moment in history, NSF centers represent the ongoing pursuit of "engineering moonshots" that propel our nation's progress in science, engineering, and innovation.

[78] "R1" is a classification in the Carnegie Classification of Institutions of Higher Education denoting an academic institution that awards doctoral degrees and has a "very high research activity" level (https://carnegieclassifications.acenet.edu/carnegie-classification/classification-methodology/basic-classification/).

NSF CONTRIBUTIONS TO INTERNET ADVANCEMENTS

It is hard to imagine a world without the internet, even for those who lived before it came into existence. The worldwide web of information—supported by a backbone of hardware and software engineering innovations—pervades almost every aspect of life, and its origins provide a case study in how early investment in research by the U.S. federal government can catalyze the growth of not just a field or an industry, but fundamental changes in society.

In 1987 the National Research Council of the National Academy of Sciences released a wide-ranging report titled *Directions in Engineering Research: An Assessment of Opportunities and Needs*. The report's authors were tasked with identifying especially important or emerging areas of engineering research and, as part of that responsibility, offered some prescient examples of how that work could "bring large and rapid improvements in the quality and diversity of everyday life":

- "Information utilities" could provide low-cost access from home or office to extensive information on virtually any subject.
- Data collection and recordkeeping could become so systematized and coordinated among institutions and consumers that most ordering, billing, and banking transactions would be done instantaneously via electronics. (NRC, 1987; pp. 18–19).

These were grand aspirations at the time, but the foundation for their realization was already being laid.[79] Two years earlier, NSF had taken over lead federal government responsibility for coordinating the development of an internetwork[80] of civilian hubs of computer resources from the Department of Defense (DoD) (Gould, 1990). NSFNET—an outgrowth of the agency's support of the establishment of the Computer Science Network (CSNET) in 1981 (Denning et al., 1981)—provided a means for the university-based supercomputer centers of that era[81] to be quickly and freely accessed by U.S. researchers and educational institutions. NSF's goal was to build and facilitate the operation of a network at a larger scale than had previously been attempted and to put in place a decentralized management structure to oversee it. Engineering the large-scale collaborations needed to agree on the underlying construction path, operational regime, and governance structure was itself a research project of great magnitude and difficulty, one that was overseen by Stephen Wolff, then the agency's Division Director for Networking and Communications Research and Infrastructure (Internet Society, n.d.).

A history of the network highlighted NSFNET's contribution to the later development of the modern internet:

It was the first large-scale implementation of internet technologies in a complex environment of many independently operated networks. NSFNET forced the internet community to iron out technical issues arising from the rapidly increasing number of computers and address many practical details of operations, management and conformance (NSF, 2010a; p. 10).

---

[79] *Continuing Innovation in Information Technology: A Workshop Report* (NASEM, 2016) includes a more detailed history of the evolution of the internet in Chapter 2.

[80] The descriptive term "internetwork" was first shorted to "internet" in the 1974 publication *Request for Comments 675 – Specification of Internet Transmission Control Program* (Cerf et al., 1974) and later came into general usage.

[81] The supercomputers centers themselves are an impact that was established with funding that was provided in part by NSF (Bement et al., 1995); they are addressed elsewhere in the report.

As networks proliferated and commercial entities took a greater interest in them, NSF identified the need to transition their stewardship to the private sector. A solicitation delineating the elements of the system that would allow the internet to grow and sustain itself was published in 1993[82], and contracts awarded in 1995 allowed the decommissioning of the NSFNET backbone that year (Frazer, 1996). Over the time period when NSF was most active in the effort—1985–1993—the number of internet-connected computers grew from 2,000 to more than 2 million (NSF, 2000d).

These early investments were not the only components of NSF's broad-ranging support that helped to make the modern internet possible. Other internet-related innovations include:

- **the widespread adoption of Transmission Control Protocol/Internet Protocol (TCP/IP)**, the set of rules that govern how computers communicate with one another over a network. Vinton Cert and Robert Kahn designed the TCP/IP protocol suite in 1970s. Dennis Jennings, the first Program Director for networking at NSF, championed its adoption as the standard for all computers and networks connected to NSFNET in 1985 (TechArchives, 2015). While there was some initial resistance to the idea of mandating a single communications protocol for all applications, it was soon recognized that such standardization was essential to the establishment of a universally accessible web (Leiner et al., 1997).

- **Digital Subscriber Line (DSL) broadband access**. Dr. John Cioffi, recognized as the father of DSL, was funded by an NSF Presidential Young Investigator[83] (in 1987) and later awards,[84] during the time in which he and his collaborators developed the methodology that underlies the operation of the modems that connect remote computers to one another via landlines (Cioffi, 2004).

- **Mosaic**, the first freely available Web browser that incorporated both graphics and text in an inline format. Mosaic was developed and refined at the NSF-funded National Center for Supercomputing Applications at the University of Illinois Urbana–Champaign. The browser, which was initially made public in 1993, allowed text, graphics, sound, and videos to co-exist on a single web page and paved the way for the multimedia browsers currently in use (History of Domains, 2020).

- **Google**. NSF was lead agency for the federal Digital Library Initiative, which made its first grants in 1994. One of these went to Stanford University[85] and supported the work of then-graduate student Larry Page. Sergey Brin, who held an NSF Graduate Research Fellowship at the university, and Page authored the seminal 1998 paper that explicated PageRank, a system for identifying the relative importance of websites in content searches that culminated in Google's search service.

---

[82] Solicitation 93-52. Network Access Point Manager, Routing Arbiter, Regional Network Providers, and Very High Speed Backbone Network Services Provider for NSFNET and the NREN Program. https://www.nsf.gov/pubs/stis1993/nsf9352/nsf9352.txt (accessed February 18, 2023).

[83] Award #8657266.

[84] Awards #9203131, #9628185, and #0427711.

[85] Award #9411306.

Another of these innovations targeted at K-12 teachers, the California Education and Research Federation network (CERFnet), was addressed in the committee's August 2022 information-gathering symposium (NASEM, 2023) in a presentation by its founder, Susan Estrada. Ms. Estrada established the network in 1986 with a $2.6 million grant from NSF. It initially connected 33 academic institutions and two commercial users (SAIC and Northrop Corporation) but quickly expanded to over 100 members. CERFnet later joined with Virginia-based networks PSInet and UUnet to form the Commercial Internet eXchange, a consortium that facilitated the implementation of e-commerce (Kende, 2000). The network also pioneered connecting K–12 teachers to the web—providing them with email access through the Free Education Mail gateway whose establishment was funded by NSF (Christianson and Fajen, 1993)—and supplying remote digital access to library databases (NRC, 1995). CERFnet's outreach activities included a comic book series titled *The Adventures of Captain Internet and CERF Boy* that sought to teach the principles of internet operation, nomenclature, and

FIGURE 4-4 The cover of volume 1 of *The Adventures of Captain Internet and CERF Boy*. SOURCE: CERFnet (1991).

use to the general public through the punning escapades of the titular characters[86] (Figure 4-4). The network has been informally credited as one of the originators of the expression "surfing the web".[87] Ms. Estrada's contributions were recognized in her 2014 induction into the Internet Hall of Fame.[88]

To be clear, none of the innovations cited here came about solely as the result of NSF funding, and other agencies of the federal government—notably the Department of Defense, Department of Energy, and NASA—played key parts in them. Later, commercial firms, some of which were started by entrepreneurs who had previously received agency funding, advanced these developments to create the myriad hardware and software components of today's web. Still, NSF support at critical junctures is widely credited with having a formative role in developing the internet of today.

The influence of these innovations on the U.S. economy and society in general is vast and has been the topic of extensive scholarship (e.g., DiMaggio et al., 2001). A detailed examination of the internet's impacts is beyond the scope of this study, but a few examples illustrate its reach:

It was estimated that the internet sector was the fourth largest sector of the U.S. economy in 2018, contributing $2.1 trillion to the nation's gross domestic product, creating 6 million direct jobs, and supporting an additional 13.1 million jobs in other areas of the economy (Hooton, 2019).

While the Federal Communications Commission has noted marked improvement in the number of people who have access to high-speed internet services and in the penetration of those services into rural areas in recent years, 14.5 million Americans still lacked such access in 2019,

---

[86] See, for example, Volume 1, October 1991, *The LAN that Time Forgot*, which is preserved on the Internet Archive (https://archive.org/details/CaptainInternetAndCERFBoyNumber1October1991).

[87] *The Women who Coined the Expression "Surfing the Internet."* (n.d.). https://www.surfertoday.com/surfing/the-woman-who-coined-the-expression-surfing-the-internet (accessed February 20,2023).

[88] https://www.internethalloffame.org/inductee/susan-estrada/.

and tribal areas continue to lag behind in deployment (FCC, 2021). Disparities in access have material effects on peoples' welfare, a circumstance exacerbated by the COVID-19 pandemic. Survey data collected in 2021 found that subjective well-being was higher during the pandemic for people with better home internet service after controlling for age, employment status, earnings, working arrangements, and other factors (Barrero et al., 2021).

The internet has transformed all aspects of self-care and medicine. Web-connected activity-tracking smartwatches, nutrition coaching and other self-improvement apps, and prescription fulfillment services help people live healthier lives, while e-Health services provide 24/7 access to authoritative advice without a person having to leave home (Lee et al., 2023). However, there are economic and social inequities in access to such amenities (Latulippe et al., 2017), and health misinformation spread via social media and online forums "has considerably harmed the adoption of recommended prevention and control behaviors and has decreased support for vital policies, such as vaccination" (Kington et al., 2021; p. 2).

Analyses presented in a series of reports by committees of the National Academies organized by its Computer Science and Telecommunications Board (NRC, 1995, 2002, 2003, 2009; NASEM 2012, 2016, 2020) document how the internet and, more broadly, innovations in wireless and broadband technologies affect and are affected by sectors as diverse as agriculture, manufacturing, and entertainment, as well as a wide range of public entities and commercial ventures.

The scope and magnitude of the engineering innovations mentioned here illustrate why "the internet" is cited in NSF's Nifty 50,[89] Sensational 60,[90] and 70th anniversary (History Wall[91]) compilations of scientific and technical developments that the agency has contributed to, and why the committee chose to highlight it as an exemplary impact.

SEMICONDUCTORS AND INTEGRATED CIRCUITS

The evolution of computing devices traces a fascinating journey from mechanical gears to vacuum tubes and transistors, each stage marking a leap in speed and efficiency. However, it was the groundbreaking invention of the integrated circuit in the early 1960s that ignited a revolution. This compact marvel, which integrates multiple semiconductor devices on a single chip, has become the driving force behind the electronic landscape that powers the modern world, be it the internet, smartphones, wireless communication including Wi-Fi and cellular, or the ongoing artificial intelligence revolution. Semiconductors have become so indispensable to our modern lives that the CHIPS and Science Act (Public Law 117–167; 2022) dedicated $52.7 billion to enhance American research, development, and manufacturing of semiconductors as well as develop a skilled semiconductor workforce,[92] thereby bolstering national defense and global competitiveness.

In the nascent era of integrated circuits, only a handful of transistors could fit on an integrated circuit chip. Contemporary integrated circuits, by contrast, contain tens of billions of transistors. This technological growth over the past six decades can be attributed to two concurrent revolutions that have significantly shaped the trajectory of modern electronics. The

---

[89] https://www.nsf.gov/about/history/nifty50/index.jsp.
[90] https://www.nsf.gov/about/history/sensational60.pdf.
[91] https://www.nsf.gov/about/history/history-wall.jsp.
[92] The act also authorized the establishment of the Technology, Innovation, and Partnerships (TIP) directorate at NSF, which is charged with advancing U.S. competitiveness through investments in key technologies including semiconductors.

first of these is the semiconductor manufacturing revolution, commonly encapsulated in "Moore's Law." This principle describes the consistent reduction in feature size—denoting the dimensions of a transistor—that has enabled the number of transistors on a microchip to double approximately every 2 years, leading to an exponential advancement in computational capabilities. Progress in digital electronics, such as the decline in microprocessor costs, expansion of memory capacity (random access memory [RAM] and flash), enhancement of sensors, and advances in digital camera pixel count and size, are closely intertwined with Moore's law.

Concurrently, an equally pivotal revolution unfolded, rooted in the concept of "abstraction." The first integrated circuits were used to create basic logic elements, such as inverters and NAND gates. The outcome of this abstraction was that engineers designing these integrated circuits did not have to design at a lower specification level and instead could provide higher-level specifications which make the design process faster and more accessible. This abstraction revolution continues today, and various milestones include generations of microprocessors, communications chips, graphics processors, and the emergence of chips enabling the artificial intelligence revolution.

These dual revolutions are a quintessential example of synergy and collaboration between industry, academia, and the Federal Government's research-funding organizations, in particular, (D)ARPA and the NSF. This discussion spotlights some of the integral contributions in integrated circuits development that have resulted from the cooperative efforts of these key stakeholders.

Mead-Conway Very Large Scale Integration (VLSI) Design

In the 1970's, integrated circuits provided a foundational set of abstractions for engineers designing electronic systems. However, delving into the design of these circuits remained a specialized art, requiring intricate knowledge of the manufacturing process. Two researchers—one from industry and one from academia—revolutionized this process. Carver Mead, a Caltech professor, taught the course "Semiconductor Devices" in 1971. After a presentation about the course at Xerox in 1976, Mead met Lynn Conway, a then Xerox PARC computer system architect, and the two co-authored the textbook *Introduction to VLSI [Very Large Scale Integration] Systems* in 1978 which revolutionized the field. VLSI is a set of tools which included the concept of "Lambda," drastically simplifying the semiconductor design process. Lambda is a unit of measurement used to express the physical dimensions of components on a chip. It is used to abstract the physical dimensions, allowing designers to work at a higher level and focus on the functionality of the components rather than their specific physical details. This breakthrough made integrated circuit design accessible to a broader range of computer scientists and engineers, particularly within the university research community. Even before Mead and Conway's book was published in 1978, the circulation of early preprint chapters in classes and among other researchers attracted widespread interest and created a community of people interested in the approach (Perkins, 2023). Very quickly, NSF began receiving proposals to fund VLSI research, and the first VLSI grant was awarded in 1978.[93] Carver's colleague, Ivan Sutherland—professor and co-founder of Caltech's Computer Science department—

---

[93] Award #7805776.

implemented VLSI courses at the heart of the program and also received funding from NSF to design with VLSI.[94]

Mead and Conway "provided the structure for a new integrated system design culture that made VLSI design both feasible and practical" (Electronics Design, 2002; p. 220); however, it would not have been possible to take advantage of this innovation without the concurrent capability to fabricate physical chips. Here is where government research and development became indispensable. A system called MOSIS, the Metal Oxide Semiconductor Implementation Service, allowed researchers to fabricate small quantities of chip prototypes using their designs at a reduced cost through a technique known as a multi-project chip, akin to the contemporary concept of "ridesharing." Originally supported by DARPA, access to MOSIS was limited until 1982 when the project was transferred to NSF. This transition broadened and democratized access to MOSIS, including NSF-sponsored researchers and institutions, making chip prototyping available to a much wider community and nurturing the technology until it reached it reached commercial viability (NRC, 1999, p. 121).

The Mead-Conway collaboration and MOSIS also helped to facilitate something even more revolutionary: the "fabless" semiconductor industry. Instead of investing in expensive fabrication facilities (fabs), fabless semiconductor companies can outsource the production of their designs to third-party semiconductor foundries. Companies can then focus on designing and marketing semiconductor chips without owning and operating the fabs. This ability to separate the design of an integrated circuit from the details of the semiconductor fabrication process fostered an explosion in the number of private companies and individuals engaged in designing integrated circuits, which includes many of the largest chip companies in the world, such as NVIDIA, Broadcom, and Qualcomm. On the foundry side, Morris Chang founded Taiwan Semiconductor Manufacturing Company (TSMC) in 1987, the first foundry dedicated solely to semiconductors. TSMC remains the largest semiconductor producer in the world, generating an estimated 90% of super-advanced semiconductor chips for fabless semiconductor companies like those listed previously (Cheung and Ripley, 2024).

Computer-Aided Design Tools Applied to Electronic Circuits

In the early years of integrated circuit design, the creation of masks used for patterning the layers of integrated circuits was a painstaking, manual process. In addition, the tools available for checking these layouts and simulating the circuits before going into fabrication were still in their infancy. The Mead–Conway revolution unleashed creativity in a whole other dimension, namely, software tools for creating, simulating, and checking the designs.

In no small part the chip industry owes its existence to the simultaneous development of sophisticated software tools. As semiconductor design automation became essential, NSF and DARPA responded to the growing complexity of chip design by supporting research that laid the groundwork for the successful commercial development of Computer Aided Design (CAD) software. By automating these activities, the ability to "tape out" a virtual design and have confidence that it would be functional when fabricated increased dramatically. In present day, it is not uncommon for multi-billion transistor chips to function seamlessly on the first fabrication attempt, or "first silicon."

One of the most crucial hardware advancements since the 1970s is a process known as "logic synthesis." Despite Mead–Conway's simplification of chip design details, engineers still

---

[94] Award #7826367.

had to design at a relatively low level—the "transistor" or "gate" level. To enhance abstraction, they began using "cell libraries," similar to the components of previous integrated circuits. Going further, engineers developed the ability to write "code" that could be "compiled" into a functional chip. Specialized languages, such as Verilog and VHDL, were created for digital system design, known as register transfer languages (RTLs). RTLs proved useful because once a design was captured in code, various tools such as simulators could be employed for higher-level validation. These simulations are critical for allowing designers to test complex integrated circuit designs before reducing them down to a "transistor" or "gate" level. RTLs made hardware design more like software design, essentially turning it into a process akin to programming. The commercialization of logic synthesis by companies such as Synopsis and Cadence marked a revolution essential to the existence of the modern computer industry.

NSF supported a number of the early contributions in the semiconductor field and the researchers responsible for them. These include:

- In the late 1960s/early 1970's, Donald Pederson and his students at the University of California, Berkeley (including Lawrence Nagel) developed Simulation Program with Integrated Circuit Emphasis (SPICE), the first universally applicable circuit simulation program, and made it widely available to industry[95] (Roessner et al., 1998).
- As an NSF-funded graduate student at MIT, Ivan Sutherland (the aforementioned co-founder of Caltech's Computer Science department) developed the first program ever to use a graphical user interface, forever changing human–computer interaction. Considered the first CAD system, "Sketchpad" was debuted in Sutherland's 1963 doctoral thesis (Cardoso, 2017). For his work he won the Turing Award in 1988 (ACM, n.d.).
- At Carnegie Mellon University (CMU), Daniel Siewiorek and colleagues received extensive NSF support for breakthroughs in digital synthesis and multiprocessor architecture. In a 1991 interview (Siewiorek, 1991), Siewiorek said that the highlights of his NSF-funded work were: Cm ("CM Star"), ISP (instruction set processor), and Micon (microprocessor configurator).
  - o Cm: A multiprocessor system developed in the early 1970s designed to coordinate large numbers of microprocessors into a modular and interconnected computing system.
  - o ISP: A computer language also developed in the 1970s, serving as a behavioral language to concisely describe computer systems at the register transfer level, with an emphasis on the automatic synthesis of high-level structures using register transfer modules. ISP is considered a predecessor to Verilog.
  - o Micon: A specialized CAD system developed in the 1980s, primarily focused on designing microprocessor-based systems for customized single-board computer design. It featured knowledge acquisition at the front, end giving users the ability to specific design features based on a series of prompts.
- Stephen Director was an early innovator in CAD with NSF CAD grants dating back to 1976 (NSF, n.d.-e). Director developed methods for maximizing the yield

---

[95] Award #GK-17931.

in integrated circuit manufacturing, leading to the "design for manufacturability" industry (Martin, 2016). In 1982, Director founded the Semiconductor Research Corporation–CMU Center for Computer-Aided Design.

The list of semiconductor technologies and the innovators driving this global paradigm shift is extensive. Among these innovators is Tsu-Jae King Liu. Liu, the child of parents who came from Taiwan to study in the United States, has made significant contributions to technology in both academia and the semiconductor industry, achievements that were facilitated in part by NSF Faculty Early Career Development Program (CAREER) funding. At Berkeley, she leads research focusing on novel semiconductor and non-volatile memory devices as well as M/NEMS technology for ultra-low power circuits. Her work is part of the Berkeley Emerging Technologies Research Center and the NSF Center for Energy Efficient Electronics in Science. Liu's accomplishments include pioneering work on polycrystalline silicon–germanium thin film technology and co-inventing—along with Chenming Hu and Jeffrey Boker—the three-dimensional fin field-effect transistor (FinFET) design, now ubiquitous in microprocessor chips (O'Reagan and Fleming, 2018). As dean of Berkeley's College of Engineering, she has been a vigorous advocate for diversity and inclusion in the field.

Organic Semiconductors

While research on traditional semiconductors such as silicon and germanium, was focused on their electrical properties, thermal conductivity, fabrication, and rigidity, in the 1950s and 1960s physicists and chemists were exploring organic semiconductors, which are made of polymers of carbon and hydrogen and which were not just available in a crystalline form but also as amorphous thin film. In 1960, while working at New York University, Harmutt Kallmann and Martin Pope discovered that organic semiconductors are electrical insulators but become semiconducting when charges are injected from electrodes. This discovery paved the way for applying organic solids as active elements of integrated circuits in today's electronic devices, such as organic light-emitting diodes, organic solar cells, and organic field-effect transistors, (OFET) now ubiquitous in high-end televisions and phones.

In 1963, Pope, a child of Ukrainian immigrants who had changed his name from Isidore Poppick due to concerns regarding anti-Jewish bias, authored a paper titled "Electroluminescence in Organic Crystals." This paper documented his findings that electricity could induce light emission from anthracene (Hafner, 2022). His research, partially supported by NSF in 1974,[96] delved into the properties of organic compounds such as anthracene and tetracene. Pope's investigations revealed that these substances possessed the essential characteristics required for the development of carbon-based electronic devices, mirroring the functionality of silicon-based counterparts. Unlike silicon, which is derived from minerals, carbon-based materials exhibiting semiconductor properties offer a distinct advantage as they can be malleable and flexible, enabling easier fabrication into thin films used in electronic applications.

This flexibility and versatility of organic semiconductors found an application in the field of biosensors using OFETs. One engineer who explored this area is Stanford's Dr. Zhenan Bao[97] whose research focuses on the design of organic electronic materials such as integrated circuits

---

[96] Award #7404764.
[97] Elected to the NAE in 2016.

with skin-like properties (flexible, stretchable, self-healing, and biodegradable). Dr. Bao hypothesized that such soft electronic sensors with skin-like properties could fundamentally change the way humans interact with electronic devices; since our health is regulated by both electrical and chemical signaling, the ability to monitor such information precisely throughout the entire body instead of at a single location should enable a significantly better understanding of human health and eventually lead to effective interventions (Materials Today, 2024). Dr. Bao has received support from NSF for fabrication[98] and patterning[99] of semiconductor devices. This work has exciting interdisciplinary applications, such as improved organic solar cell efficiency,[100] AI-enabled multimodal stress detection for precision medicine,[101] and multi-layer self-healing synthetic electronic bandages for artificial skin on prosthetic limbs. Through her co-founded startup companies, Bao has helped market conductive ink (C3 Nano, 2015) and wireless non-invasive blood pressure monitoring (PyrAmes, 2024).

Observations

Computing devices have undergone a remarkable journey, with integrated circuits emerging as a transformative force in shaping the modern electronic landscape. From Mead–Conway's revolutionary Very Large Scale Integration to the democratization of chip prototyping through NSF's MOSIS and the advent of the fabless semiconductor industry, collaborative efforts have propelled the field forward. The development of Computer Aided Design tools applied to electronic circuits further accelerated progress, automating complex tasks and enabling nearly seamless functionality of multi-billion transistor chips. The intertwining of hardware and software advancements, exemplified by logic synthesis and register transfer languages, has turned hardware design into a process akin to programming.

Acknowledging the pivotal role played by individuals such as Tsu-Jae King Liu, who has not only made significant contributions to semiconductor technology but also advocated for diversity and inclusion in engineering, underscores the holistic nature of this technological paradigm shift. Reflecting on these parallel revolutions, it becomes evident that the collaborative efforts of industry, academia, and government organizations including the NSF have been instrumental in propelling the United States and the world into the era of advanced computing and shaping the trajectory for the 21st century.

WIND ENERGY TECHNOLOGY

Energy, indispensable for our quality of life, serves critical functions in lighting, heating, and cooling homes and facilitating transportation. As a fundamental component of industrial processes, it plays a vital role in global infrastructure. Due to global population growth and development, between now and 2060 the global primary energy demand is projected to increase by one-third and the proportion of energy from renewable sources versus fossil fuels is expected to double or triple (Kober et al., 2020). Addressing this demand while mitigating climate change requires a shift toward diverse energy sources, including renewables, biofuels, electricity, hydrogen, and gas.

---

[98] Award #1006989.
[99] Award #0705687.
[100] Award #1434799.
[101] Award #2037304.

While most renewable energy -related research in the U.S. is funded at the federal level by the Department of Energy, NSF's history of investments has also advanced knowledge and created economic opportunities in some specific areas. These include agency support of fundamental and applied research on solar panels, leading to improvements in design and efficiency and increased solar energy production (NSF, n.d-j); more efficient and sustainable means of manufacturing biofuels (ITIF, 2024), and innovations in wind energy technology. The last of these is the focus of this section.

Wind energy, a centuries-old renewable source, has become a critical contributor to the U.S. energy portfolio. Constituting only 1 percent of total electricity generation in the United States in 1990, wind power had increased to approximately 10.3 percent by 2020 (EIA, 2023a,b). The Department of Energy (DOE) has set a goal of 20 percent wind power by 2030 (DOE, 2008), including the Biden–Harris administration's goal of 30 gigawatts of offshore wind power by 2030—sufficient to power 10 million homes with purely offshore electricity (DOI, 2022). The roots of wind energy research in the United States trace back to the oil crisis of 1973, prompting the establishment of the Energy Research and Development Administration (ERDA) and investment in renewable energy (Anders, 1980). In collaboration with the NSF and private partners, ERDA funded NASA's Lewis Research Center in Sandusky, Ohio, where a prototype wind turbine—the 100-kW Mod-0 with three blades—was developed in 1975, laying the foundation for the contemporary wind turbine industry.

As in the development of the 100-kW Mod-0, government agencies have played complementary roles in the advance of wind energy and renewable energy overall. This is exemplified by the memorandum of understanding signed in 2022 between NSF and DOE's Office of Energy Efficiency and Renewable Energy, affirming the continuation of their longstanding collaboration and the shared commitment to decarbonize the U.S. economy by 2050 (NSF, 2022e). While the DOE has historically specialized in applied research and development, translating foundational discoveries into scalable solutions, NSF usually contributes by advancing foundational research, funding academic institutions and research facilities, and supporting cutting-edge research for technological breakthroughs (NSF, 2024). However, in recent years NSF has taken a more active role in supporting applied research as well.

One example is the NSF-funded Industry–University Cooperative Research Center (IUCRC) for Wind Energy, Science, Technology and Research (WindSTAR), established in 2014 through a collaboration between the University of Massachusetts (UMass) Lowell and the University of Texas at Dallas (UTD). WindSTAR focuses on addressing the critical needs of the wind industry (WindSTAR, 2023), thus facilitating multi-sectoral collaboration between university and industry partners to improve the economic and technical viability of wind power at all stages of wind power plant development. Industry partners include wind turbine manufacturers, producers of crucial wind turbine components, suppliers of specialized equipment and consultants, service providers, and wind project developers along with associated industries (NSF, n.d.-b). As of 2020, WindSTAR had carried out 50 research projects of benefit to their industry partners (NSF, 2020b), securing $3.9 million in funding between 2014 and 2019, with an additional $1.1 million from NSF to sustain operations until 2024 (UTD, n.d.-b).

At UMass, the focus is on advancing materials, manufacturing, and testing of wind turbines as well as on exploring energy storage, transmission, and zero-carbon fuel generation. Meanwhile, UTD specializes in high-fidelity simulations of wind power systems, using LiDAR measurements and wind tunnel testing to design efficient turbines. UTD also contributes to training the next generation of renewable energy professionals through programs such as NSF's Research Experience for Undergraduates (REU)[102] and Non-Academic Research Internships for Graduate Students (INTERN).[103] Collaborative research efforts include optimizing wind blade repairs, predicting turbine joint lifetimes, using machine learning to control blade manufacturing processes, and developing structural health monitoring systems for turbine blades (WindSTAR, 2023).

Other significant wind energy projects supported by NSF include:

- **Simulating wind power expansion with supercomputing**: A Cornell University study used advanced supercomputer simulations to model plausible scenarios for expanding U.S. wind power capacity. The research suggested the potential to double or quadruple wind power capacity, emphasizing the use of next-generation, larger wind turbines to minimize impacts on system-wide efficiency and local climate in order to meet the DOE target of achieving 20 percent wind power electricity by 2030 without requiring additional land (NSF, 2020c).
- **Centers of Research Excellence in Science and Technology (CREST) Center for Energy and Environmental Sustainability (CEES—Phases I and II)**: Located at Prairie View A&M University, a Historically Black University, this center has made significant contributions to the renewable energy sector including wind energy[104]. Research achievements include understanding flow dynamics to enhance efficiency through pointed-tip turbine blades, contributing to a potential power increase in the dynamic stall region (10–15 m/s wind speed) compared with the original National Renewable Energy Laboratory blade. Additionally, CEES plays a vital role in education and outreach, establishing an energy engineering minor, awarding scholarships, and engaging with the Greater Houston community through publications, presentations, and community initiatives.
- **U.S./Europe Partnership for International Research and Education (PIRE) in Wind Energy Intermittency (WINDINSPIRE)**: Led by Johns Hopkins University in partnership with U.S. and European research partners, WINDINSPIRE addressed crucial research questions related to integrating intermittent wind sources into power systems[105]. The project developed improved tools for computational fluid dynamics modeling of wind farms, enhancing predictions of power fluctuations and creating dynamic models for wind farm controls. WINDINSPIRE also contributed to the development of methods for estimating spatio-temporal variability in power output, models for efficient resource allocation in grid planning, and tools for analyzing market designs and regulatory choices. Beyond its research impact, WINDINSPIRE played a significant role in educating and training the next generation of wind researchers through international collaborations, symposia, and research experiences for graduate and undergraduate students.

The fundamental research driving advances in the wind industry is spearheaded by leaders such as Dr. John Oluseun Dabiri, a Nigerian–American aeronautics engineer and

---

[102] Award # 2150488.
[103] Award # 1362033.
[104] Awards #1036593 and #1914692.
[105] Award #1243482.

Centennial Chair Professor at the California Institute of Technology. His research as director of the Biological Propulsion Laboratory (Center for Biologically Inspired Design, 2014) has delved into fluid transport and flow dynamics in aquatic locomotion, energy conversion, and cardiac flows. This work includes the design of a vertical-axis wind farm inspired by the formations of schooling fish,[106] whose coordinated movements produce wakes that aid in the motion of surrounding fish. Taking advantage of analogous physical properties in adjacent vertical-axis wind turbines can significantly enhance wind farm power production.

Dabiri's innovative approach to wind farm design involves closely spaced pairs of counter-rotating turbines that efficiently funnel air to neighboring turbines, minimizing energy loss to turbulence (Dabiri, 2011; Pritchard, 2011). This design not only benefits adjacent turbines but also demonstrates a power generation boost compared with turbines working independently. In tests, turbines positioned five rows back still generated 95 percent of the power of those in the front row. This closely packed wind farm design has the potential to produce 20 to 30 watts per square meter of land, approximately ten times the output of current wind farms. Dabiri and some of his students implemented the experimental wind farm design in the indigenous community of Igiugig, Alaska, supported by a grant from the Gordon and Betty Moore Foundation.[107] The turbines help to power a rural community that had traditionally relied on more expensive diesel generators. One of Dabiri's students created a startup, XFlow,[108] to continue the project.

Dabiri's important research contributions in a variety of fields have garnered recognition, including a MacArthur Fellowship, a Presidential Early Career Award for Scientists and Engineers (PECASE), and his service on the President's Council of Advisors on Science and Technology (PCAST), along with receiving the 2020 NSF Alan T. Waterman Award, recognizing outstanding young science and engineering researchers.

Another exemplary figure with significant contributions to both academia and the field of wind energy is Dr. Mario Rotea. As the Erik Jonsson Chair in Engineering and Computer Science at the University of Texas at Dallas (UTD), he has played a pivotal role in advancing wind energy science and engineering. Rotea co-founded "WindSTAR," the aforementioned NSF Industry—University Cooperative Research Center that fosters collaboration between academia and industry to drive forward wind energy through relevant research. Additionally, he serves as the director of "UTD Wind," a center specifically established for the advancement of wind energy science (UTD, n.d.-a).

Rotea has held various leadership positions, including heading the mechanical engineering department at UTD, where he oversaw significant growth in student enrollment, faculty size, and the establishment of a Ph.D. program (NSF, 2015a). Prior to his tenure at UTD, he significantly expanded the Mechanical and Industrial Engineering Department at the University of Massachusetts Amherst, particularly in the area of wind energy and industrial engineering applications in healthcare. Rotea's commitment to interdisciplinary collaboration is evident in his co-directorship of WindSTAR, bringing together researchers from UTD and the University of Massachusetts Lowell with industry partners. His work demonstrates a dedication to both research and education, aligning with the NSF's mission as seen in his terms as director of the Control Systems Program and division director of Engineering Education and Centers at NSF. Rotea's influence extends beyond academia, as he worked for the United Technologies

---

[106] Award #0725164.
[107] https://www.uaa.alaska.edu/news/archive/2015/09/harnessing-the-wind-in-igiugig-for-village-sized-energy-alternatives.cshtml.
[108] https://www.xflowenergy.com/.

Research Center, contributing to the development of advanced control systems for various applications, including helicopters, gas turbines, and machine tools. Recognized as a Fellow of the Institute of Electrical and Electronics Engineers (IEEE), his contributions to robust and optimal control of multivariable systems have significantly shaped the field.

Looking towards 2050, anticipating a global population increase to nearly 10 billion people (UN, 2022) and electricity needs that are predicted to more than double by 2060 (World Energy Council, 2019), the imperative for a low-carbon energy future becomes ever more pronounced. The dual challenge facing society involves extending energy benefits to all while effectively managing the risks of climate change. This energy transition, under way globally, will unfold at varying paces and yield different outcomes based on local factors such as available natural resources, weather patterns, national climate change policies, economic growth, and the adoption of technologies. To address this challenge, collaboration among policymakers, private industry, governmental agencies, and academic institutions is crucial. In the United States, NSF is a key player in this collaborative effort. The wind energy sector is an example of how multidisciplinary and multisectoral collaborations, funding academic institutions, facilitating cutting-edge research, and supporting innovative researchers contributes to the ambitious goal of transitioning to a sustainable and efficient energy future.

CONCLUSIONS

As the committee's research—and, in particular, the descriptions of the exemplary impacts presented in this chapter—make clear, NSF funding of engineering research and education has had profound societal effects. On the basis of these descriptions and the additional information presented in this and earlier chapters of the report, the committee has reached the following conclusions.

- NSF's investments in engineering education and research have played a catalytic role in advancing the science, technology, and engineering ecosystems.
- NSF's support of interdisciplinary and intersectoral collaboration on research initiatives and of centers has contributed to engineering's positive impacts on society.
- NSF investments in women and in others underrepresented in the engineering field and in fostering a more supportive learning and research environment for these groups have had a part in bringing about significant engineering contributions.

REFERENCES

ACM (Association for Computing Machinery). n.d. *Ivan Sutherland.*
    https://awards.acm.org/award-recipients/sutherland_3467412 (accessed March 1, 2024).
Agarwal, R. 2022. The personal protective equipment fabricated via 3D printing technology
    during COVID-19. Annals of 3D Printed Medicine 5: 100042.
    https://doi.org/10.1016/j.stlm.2021.100042.
AISES (American Indian Science and Engineering Society). 2016. *About AISES*
    https://www.aises.org/about (accessed March 1, 2024).
AISES. 2022. *Sarah EchoHawk.* https://www.aises.org/about/staff/sarah-echohawk (accessed
    April 21, 2024).

Anders, R. 1980. *The Federal Energy Administration.* U.S. DOE.
    https://www.energy.gov/management/articles/federal-energy-administration. (accessed March
    21, 2024).

Armstrong, Z. 2022. *Nanotechnology startup opens facility in Atlanta's West End.*
    https://www.bizjournals.com/atlanta/inno/stories/news/2022/09/02/nanotechnology-startup-
    opens-facility-in-west-end.html (accessed February 29, 2024).

Ascend Elements. 2023. *Ascend Elements opens North America's largest electric vehicle battery
    recycling facility in Georgia.* https://www.prnewswire.com/news-releases/ascend-elements-
    opens-north-americas-largest-electric-vehicle-battery-recycling-facility-in-georgia-
    301786245.html (accessed February 29, 2024).

ASTRO America. 2022. *AM Forward.* https://astroa.org/project/am-forward/ (accessed February
    28, 2024).

Baker, W. 1987. Advances in materials research and development. In National Research Council,
    Advancing materials research. Washington, DC: National Academy Press.
    https://doi.org/10.17226/10291. Pp. 3–24.

Barrero, J. M., N. Bloom, and S. J. Davis. 2021. *Internet access and its implications for
    productivity, inequality, and resilience.* National Bureau of Economic Research working
    paper 29102. https://www.nber.org/papers/w29102

Bement, A. L. Jr., P. A. Kollman, M. K. Vernon, J. Hennessy, A. B. White Jr., J. Ingram, A.
    Schlumberger, W. A. Wulf, N. Pitts, R. Voigt, and P. R. Young. 1995. *Report of the Task
    Force on the Future of the NSF Supercomputer Centers Program.*
    https://nsf.gov/pubs/1996/nsf9646/nsf9646.pdf (accessed February 18, 2023).

Bennett, J., C. Elkan, and B. Liu. 2007. KDD Cup and Workshop 2007. *SIGKDD Explorations*
    9(2): 51–52.

Biomedical Engineering Society. 2004. *BMES: Celebrating 35 years of biomedical engineering:
    An historical perspective.* Biomedical Engineering Society. https://assets.noviams.com/novi-
    file-uploads/bmes/PDFs_and_Documents/History/BMES_35_year_HISTORY_-
    _FINAL_1_.pdf (accessed April 16, 2024).

Borrego, M., R. S. Adams, J. Froyd, L. R. Lattuca, P. T. Terenzini, and B. Harper. 2007. Panel -
    emerging results: Were the engineering education coalitions an effective intervention? *2007
    37th Annual Frontiers In Education Conference - Global Engineering: Knowledge Without
    Borders, Opportunities Without Passports*: F4F-1-F4F-6.
    https://doi.org/10.1109/FIE.2007.4418188

Bozeman, B., and C. Boardman. 2004. The NSF Engineering Research Centers and the
    university–industry research revolution: A brief history featuring an interview with Erich
    Bloch. *Journal of Technology Transfer* 29(3-4):365–375.

Brandom, R. 2016. Humanity and AI will be inseparable, says CMU's head of machine learning.
    The Verge, November 15. http://www.theverge.com/a/verge-2021/humanity-and-ai-will-be-
    inseparable (accessed March 1, 2024).

Brin, S., and L. Page. 1998. The anatomy of a large-scale hypertextual web search engine.
    Computer Networks and ISDN Systems 30(1–7):107–117.

C3 Nano. 2015. *Technology overview—C3Nano.* https://c3nano.com/technology/,
    https://c3nano.com/technology/ (accessed March 1, 2024).

CAHSI (Computing Alliance of Hispanic-Serving Institutions). n.d. *CAHSI overview.*
    https://cahsi.utep.edu/about/ (accessed April 21, 2024).

Cazzaniga, M,. F. Jaumotte, L. Longji, G. Melina, A. J. Panton, C. Pizzinelli, E. J. Rockall, and M. M. Tavares. 2024. *Gen-AI: Artificial intelligence and the future of work.* https://www.imf.org/-/media/Files/Publications/SDN/2024/English/SDNEA2024001.ashx International Monetary Fund staff discussion note (accessed June 4, 2024).

Cardoso, D. 2017. *Experimental archaeology of CAD.* http://dcardo.com/projects/archaeology_of_cad/index.html (accessed April 23, 2024).

CC Corp. n.d. *Space applications.* https://www.contourcrafting.com/space (accessed February 28, 2024).

CEA (White House Council of Economic Advisers). 2022. *Using additive manufacturing to improve supply chain resilience and bolster small and mid-size firms.* https://www.whitehouse.gov/cea/written-materials/2022/05/09/using-additive-manufacturing-to-improve-supply-chain-resilience-and-bolster-small-and-mid-size-firms/ (accessed February 28, 2024).

Center for an Informed Public. 2021. *$2.25 million National Science Foundation funding will inform Center for an Informed Public's rapid response research of mis- disinformation.* Center for an Informed Public. https://www.cip.uw.edu/2021/08/15/national-science-foundation-uw-cip-misinformation-rapid-response-research/ (accessed April 16, 2024).

Center for Biologically Inspired Design. 2014. *Biological Propulsion Laboratory (Caltech) | BioInspired!* https://bioinspired.sinet.ca/content/biological-propulsion-laboratory-caltech (accessed February 29, 2024).

Cerf, V., Y. Dalal, and C. Sunshine. 1974. *Request for comments 675. Specification of internet transmission control program.* December. https://www.rfc-editor.org/rfc/rfc675 (accessed February 23, 2023).

CERFnet. 1991. *The adventures of Captain Internet and CERF Boy—Number 1.* October. https://archive.org/details/CaptainInternetAndCERFBoyNumber1October1991 (accessed March 8, 2024)

Chamot, J. 2003. Artificial retinas may restore partial sight for thousands in the next decade. The National Science Foundation. June 25. https://beta.nsf.gov/news/artificial-retinas-may-restore-partial-sight (accessed April 16, 2024).

Cheung, E., and W. Ripley. 2024, March 23. *Everyone wants the latest chips. That's causing a huge headache for the world's biggest supplier | CNN Business.* https://www.cnn.com/2024/03/22/tech/taiwan-tsmc-talent-shortage-training-center-intl-hnk/index.html (accessed May 29, 2024).

Christianson, J. S., and A. L. Fajen. 1993. Education in the matrix: The FrEdMail network. *Boardwatch Magazine* March:33,38. https://www.christiansonjs.com/wp-content/uploads/2022/01/Boardwatch_Education_In_The_Matrix.pdf (accessed February 20, 2023).

Cioffi, J. M. 2004. Ask me again in ten years! *IEEE Signal Processing Magazine* 21(6):8–11. https://ieeexplore.ieee.org/stamp/stamp.jsp?arnumber=1359137 (accessed February 20,2023).

CMU (Carnegie Mellon University). 2016. *Data Storage Systems Center.* https://www.dssc.ece.cmu.edu/ (accessed November 9, 2023).

Cohn, A. 2020. A journey to emotional (and artificial) intelligence. *Forbes*, July 24. https://www.forbes.com/sites/alisacohn/2020/07/24/a-journey-to-emotional-and-artificial-intelligence/ (accessed March 1, 2024).

Conrad, L. F., J. L. Auerbach, and A. Howard. 2015. *The impact of a robotics summer undergraduate research experience on increasing the pipeline to graduate school.* In 2015 ASEE Annual Conference and Exposition. Pp. 26.1538.1–13. https://peer.asee.org/the-impact-of-a-robotics-summer-undergraduate-research-experience-on-increasing-the-pipeline-to-graduate-school (accessed April 21, 2024).

Cranor, L. F., and S. Garfinkel. 2005. *Security and usability: Designing secure systems that people can use.* Newton, MA: O'Reilly Media.

Crevier, D. 1993. *AI: The tumultuous history of the search for artificial intelligence.* New York: Basic Books.

Dabiri, J. O. 2011. Potential order-of-magnitude enhancement of wind farm power density via counter-rotating vertical-axis wind turbine arrays. *Journal of Renewable and Sustainable Energy* 3(4):043104. https://doi.org/10.1063/1.3608170.

Davies, S. 2020. *Invention, optimism, & realism: The back story of SLS 3D printing with Dr. Joe Beaman.* https://www.tctmagazine.com/api/content/bc50c9c2-4f3a-11ea-bcf5-1244d5f7c7c6/ (accessed February 28, 2024).

DeAro, J., S. Bird, and R. S. Mitchell. 2019. NSF ADVANCE and gender equity: Past, present, and future of systemic institutional transformation strategies. *Equality, Diversity and Inclusion: An International Journal* 38(2):131–139.

Deng, J., W. Dong, R. Socher, L.-J. Li, K. Li, K., and L. Fei-Fei. 2009. ImageNet: A large-scale hierarchical image database. 2009 *IEEE Conference on Computer Vision and Pattern Recognition,* pp. 248–255. https://doi.org/10.1109/CVPR.2009.5206848

Denning, P. J., A. Hearn, and C. W. Kern. 1983. History and overview of CSNET. *ACM SIGCOMM Computer Communication Review* 13(2):138–145.

Desktop Metal. n.d. *About Desktop Metal.* https://www.desktopmetal.com/about-us (accessed February 28, 2024).

DiMaggio, P., E. Hargittai, W. R. Neuman, and J. P. Robinson. 2001. Social implications of the Internet. *Annual Review of Sociology* 27(1):307–336. https://www.jstor.org/stable/pdf/2678624 (accessed February 20, 2023).

DoD (U.S. Department of Defense). 2023. *DOD Invests $40 Million to Establish Research Centers of Excellence at Minority-serving In.* https://www.defense.gov/News/Releases/Release/Article/3560060/dod-invests-40-million-to-establish-research-centers-of-excellence-at-minority/https%3A%2F%2Fwww.defense.gov%2FNews%2FReleases%2FRelease%2FArticle%2F3560060%2Fdod-invests-40-million-to-establish-research-centers-of-excellence-at-minority%2F (accessed June 4, 2024).

DOE (U.S. Department of Energy). 2008. *20% wind energy by 2023: Increasing wind energy's contribution to U.S. electricity supply: Executive summary.* https://www.nrel.gov/docs/fy09osti/42864.pdf (accessed April 23, 2024).

DOI (U.S. Department of the Interior). 2022. *Fact sheet: Biden–Harris Administration announces new actions to expand U.S. offshore wind energy* [Press Release]. https://www.doi.gov/pressreleases/fact-sheet-biden-harris-administration-announces-new-actions-expand-us-offshore-wind (accessed February 29, 2024).

Donovan, R. 2020. *2020 annual report—The global economics of disability.* Return on Disability Group. https://www.rod-group.com/research-insights/annual-report-2020/ (accessed April 30, 2024).

EIA (U.S. Energy Information Administration). 2023a. *Frequently asked questions (FAQs)*. https://www.eia.gov/tools/faqs/faq.php (accessed February 29, 2024).

EIA. 2023b. *History of wind power*. https://www.eia.gov/energyexplained/wind/history-of-wind-power.php (accessed February 29, 2024).

Electronics Design. 2002. *Hall of fame; Honor role*. https://ai.eecs.umich.edu/people/conway/Awards/ElectronicDesign/ED%20Hall%20of%20Fame%202002.pdf (accessed June 2, 2024).

Errick, K. 2022. *NSF Invests $25.4M into Cybersecurity and Privacy Research Projects*. https://www.nextgov.com/emerging-tech/2022/08/nsf-invests-254m-cybersecurity-and-privacy-research-projects/375274/ (accessed June 6, 2024).

FCC (Federal Communications Commission). 2021. *Fourteenth broadband deployment report*. FCC 21-18. GN Docket No. 20-269. https://docs.fcc.gov/public/attachments/FCC-21-18A1.pdf (accessed February 21, 2023).

Flores, B. C., J. Shenberger-Trujillo, and M. Montes. 2022. *Mentoring graduate underrepresented minorities in STEM*. Making Connections. https://uen.pressbooks.pub/makingconnections/chapter/mentoring-graduate-underrepresented-minorities-in-stem/ (accessed March 1, 2024).

Frazer, K. D. 1996. NSFNET: *A partnership for high-speed networking: Final report, 1987–1995*. Merit Network. https://www.merit.edu/wp-content/uploads/2019/06/NSFNET_final-1.pdf (accessed February 18, 2023).

Freeman, P., W. R. Adrion, and W. Aspray. 2019. *Computing and the National Science Foundation: Building a foundation for modern computing, 1950–2016*. New York: ACM Books.

Garfinkel, S. 1998. Enter the Dragon. *Technology Review*, September 1. https://www.technologyreview.com/1998/09/01/236899/enter-the-dragon/ (accessed March 1, 2024).

Gates, A. Q. 2017. *The CAHSI INCLUDES collective impact initiative*. Pp. 67–71 in Proceeding of NSF INCLUDES Conference: Advancing the Collective Impact of Retention and Continuation Strategies for Hispanics and other Underrepresented Minorities in STEM Fields. https://par.nsf.gov/servlets/purl/10049620 (accessed April 21, 2024).

Gates, K. A. 2011. *Our biometric future: Facial recognition technology and the culture of surveillance*. New York: NYU Press. https://www.jstor.org/stable/j.ctt9qg8xd (accessed March 1, 2024).

Georgia Tech. n.d. *Baratunde A. Cola*. https://www.me.gatech.edu/faculty/cola (accessed February 29, 2024).

Georgia Tech. n.d.-a. *Ayanna Howard*. https://howard.ece.gatech.edu/ (accessed April 16, 2024).

Georgia Tech. n.d.-b. *Ayanna Howard, Ph.D.--Director of HumAnS Lab*. https://humanslab.ece.gatech.edu/people/ (accessed April 16, 2024).

Gershgorn, D. 2017, July 26. The data that transformed AI research—and possibly the world. *Quartz*, July 26. https://qz.com/1034972/the-data-that-changed-the-direction-of-ai-research-and-possibly-the-world (accessed March 1, 2024).

Gladwell, M. 1999. The science of the sleeper. *The New Yorker*, September 26. https://www.newyorker.com/magazine/1999/10/04/the-science-of-the-sleeper (accessed February 29, 2024).

Gold, J. R., A. J. Gates, S. A. Haque, M. C. Melson, L. K. Nelson, and K. Zippel. 2022. The NSF ADVANCE network of organizations. *ADVANCE Journal* 3(1):33822.

Gould, S. B. 1990. An intellectual utility for science and technology: The National Research and Education Network. *Government Information Quarterly* 7(4):415–425.

Guerini, R. 2024. Carnegie Mellon, UC San Diego top US grant winners for AI research. *Science|Business* 14 May. https://sciencebusiness.net/news/carnegie-mellon-uc-san-diego-top-us-grant-winners-ai-research (accessed June 5, 2024).

Hafner, K. 2022. Martin Pope, whose research led to OLEDs, dies at 103. *The New York Times*. March 27. https://www.nytimes.com/2022/03/27/obituaries/martin-pope-dead.html (accessed March 1, 2024).

Harvard Magazine. 2013. *Harvard bioengineering professor Jennifer Lewis prints 3-D on a micron scale.* https://www.harvardmagazine.com/node/42927 (accessed February 28, 2024).

Heimgartner, J. 2022. *Coral relief: How 3D printing is reviving the ocean's key ecosystems.* https://www.engineering.com/story/coral-relief-how-3d-printing-is-reviving-the-oceans-key-ecosystems (accessed February 28, 2024).

Hempel, J. 2018. Fei-Fei Li's quest to make machines better for humanity. *Wired*, November 13. https://www.wired.com/story/fei-fei-li-artificial-intelligence-humanity/ (accessed March 1, 2024).

Henderson, S., and M. Golden. 2015. *Self-driving cars: Mapping access to a technology revolution.* Washington, DC: National Council on Disability.

Hill, K. 2023. Santa Barbara nonprofit delivers vital telehealth care to front lines in war-torn Ukraine. *Noozhawk,* February 27. http://www.noozhawk.com/santa-barbara-nonprofit-delivers-vital-health-care-to-front-lines-in-war-torn-ukraine/ (accessed March 1, 2024).

History of Domains. 2020. *Mosaic web browser and its release.* https://www.historyofdomains.com/mosaic/ (accessed February 20, 2023).

Hooton, C. 2019. *Measuring the U.S. internet sector: 2019.* Internet Association. https://internetassociation.org/publications/measuring-us-internet-sector-2019/ (accessed February 21, 2023).

Howard, A. 2020. *Sex, Race, and Robots: How to Be Human in the Age of AI.* Audible Originals.

Huang, X., J. Baker, and R. Reddy. 2014. A historical perspective of speech recognition. *Communications of the ACM*, January 1. https://cacm.acm.org/research/a-historical-perspective-of-speech-recognition/ (accessed March 1, 2024).

Hug, S., H. Thiry, and M. McKay. 2021. *External evaluation of the CAHSI INCLUDES Alliance, 2021.* https://cahCAHSIsi.utep.edu/wp-content/uploads/cahsi-includes-evaluation-report-august-2021.pdf (accessed April 21, 2024).

Hunter College. 2014. *Alumna Mildred Dresselhaus '51 receives Presidential Medal of Freedom.* https://www.youtube.com/watch?v=T932NSNSRSE (accessed April 22, 2024).

IBM. 2023. *AI vs. Machine Learning vs. Deep Learning vs. Neural Networks | IBM.* https://www.ibm.com/think/topics/ai-vs-machine-learning-vs-deep-learning-vs-neural-networks (accessed May 29, 2024).

IBM. 2024. *Speech recognition.* https://www.ibm.com/history/voice-recognition (accessed March 1, 2024).

IGERT (Integrative Graduate Education and Research Traineeship). 2011. *Achievement: Password composition and strength.* http://www.igert.org/spotlights/2542.html (accessed May 30, 2024).

Internet Society. n.d. *A brief history of the internet.* https://www.internetsociety.org/internet/history-internet/brief-history-internet/ (accessed June 2, 2024).

ITIF (Information Technology & Innovation Foundation). 2024. *How federal funding for basic research spurs clean energy discoveries the world needs: Eight case studies.* https://itif.org/publications/2024/03/13/federal-funding-for-basic-research-spurs-clean-energy-discoveries-eight-case-studies/ (accessed June 3, 2024).

Jackson, B. 2017. *Interview with Harvard 3D printing pioneer Jennifer Lewis.* https://3dprintingindustry.com/news/interview-harvards-3d-printing-pioneer-jennifer-lewis-106645/ (accessed February 28, 2024).

Karger, P. A., and R. R. Schell. 2002. *Thirty years later: Lessons from the Multics security evaluation.* www.acsac.org/2002/papers/classic-multics.pdf (accessed April 16, 2024).

Kelso, H. 2022. *Compostable bioleather offers sustainable solutions for the clothing industry and beyond.* https://www.engineering.columbia.edu/news/compostable-bioleather-offers-sustainable-solutions-clothing-industry-and-beyond (accessed February 29, 2024).

Kende, M. 2000. *The digital handshake: Connecting internet backbones.* OPP Working Paper No. 32. Office of Plans and Policy. Federal Communications Commission. https://docs.fcc.gov/public/attachments/DOC-206157A2.pdf (accessed February 20, 2023).

Kington, R., S. Arnesen, W.-Y. S. Chou, S. Curry, D. Lazer, and A. Villarruel. 2021. *Identifying credible sources of health information in social media: Principles and attributes.* NAM Perspectives. Discussion Paper, National Academy of Medicine, Washington, DC. https://doi.org/10.31478/202107a (accessed February 23, 2023).

Kober, T., H.-W. Schiffer, M. Densing, and E. Panos, E. 2020. Global energy perspectives to 2060—WEC's World Energy Scenarios 2019. *Energy Strategy Reviews* 31:100523. https://doi.org/10.1016/j.esr.2020.100523.

Latulippe, K., C. Hamel, and D. Giroux. 2017. Social health inequalities and eHealth: A literature review with qualitative synthesis of theoretical and empirical studies. *Journal of Medical Internet Research.* 19(4):e136. https://www.jmir.org/2017/4/e136/ (accessed February 23,2023).

Lee, A., M. Corneille, K. T. Jackson, B. Banks, S. Allen, R. N. Coger, M. I. Kanipes, and S. Luster-Teasley. 2022. Narratives of Black women STEM faculty: Breaking barriers to promote institutional transformation at historically Black colleges and universities. *ADVANCE Journal* 3(1):33797. https://www.advancejournal.org/article/33797-narratives-of-black-women-stem-faculty-breaking-barriers-to-promote-institutional-transformation-at-historically-black-colleges-and-universities (accessed April 21, 2024).

Lee, P., A. Abernethy, D. Shaywitz, A. V. Gundlapalli, J. Weinstein, P. M. Doraiswamy, K. Schulman, and S. Madhavan. 2023. Digital health COVID-19 impact assessment: Lessons learned and compelling needs. In National Academy of Medicine. 2023. Emerging stronger from COVID-19: Priorities for health system transformation. Washington, DC: The National Academies Press. https://doi.org/10.17226/26657. Pp. 177–234.

Leiner, B. M., V. G. Cerf, D. D. Clark, R. E. Kahn, L. Kleinrock, D. C. Lynch, J. Postel, L. G. Roberts, and S. Wolff. *A brief history of the Internet.* 1997. The Internet Society. https://www.internetsociety.org/wp-content/uploads/2017/09/ISOC-History-of-the-Internet_1997.pdf (accessed February 20, 2023).

Lewis, C. 2010. *Engineering research centers innovations: ERC-generated commercialized products, processes, and startups.* SciTech Communications LLC. https://erc-assoc.org/sites/default/files/topics/policy_studies/ERC_innovations_2010-final.pdf. (accessed November 9, 2023).

Lush Prize. n.d. *Science prize.* https://lushprize.org/awards/science-prize/ (accessed February 28, 2024).

Martin, G. S. 2016. Provost emeritus Stephen Director named National Academy of Inventors fellow. *Northeastern Global News*, December 19. https://news.northeastern.edu/2016/12/19/provost-emeritus-stephen-director-named-national-academy-of-inventors-fellow/ (accessed March 1, 2024).

Materials Today. 2024. *Lab profile: Zhenan Bao, Stanford University.* https://www.materialstoday.com/lab-profile-zhenan-bao-stanford-university/ (accessed March 1, 2024).

McCarthy, J., M. L. Minsky, N. Rochester, and C. E. Shannon. 2006. A proposal for the Dartmouth summer research project on artificial intelligence, August 31, 1955. *AI Magazine* 27(4):4. https://doi.org/10.1609/aimag.v27i4.1904.

Meyers, C. 2021. *Karan Watson recognized with lifetime achievement award.* https://engineering.tamu.edu/news/2021/08/karan-watson-recognized-with-lifetime-achievement-award.html (accessed March 1, 2024).

MIT. n.d.-a. *Paula T. Hammond—MIT chemical engineering.* https://cheme.mit.edu/profile/paula-t-hammond/ (accessed February 29, 2024).

MIT. n.d.-b. *The Robert Morris internet worm.* https://groups.csail.mit.edu/mac/classes/6.805/articles/morris-worm.html (accessed March 1, 2024).

Morimoto, S. A. 2022. The social science of institutional transformation: Intersectional change in the academy. *Frontiers in Sociology* 7:824497.

Morimoto, S. A., A. M. Zajicek, V. H. Hunt, and R. Lisnic. 2013. Beyond binders full of women: NSF ADVANCE and initiatives for institutional transformation. *Sociological Spectrum* 33(5):397–415.

NAE (National Academy of Engineering). 1983. *Guidelines for engineering research centers.* Washington, DC: National Academy of Engineering. https://doi.org/10.17226/19472.

NAE. 2005. *Educating the engineer of 2020: Adapting engineering education to the new century.* Washington, DC: The National Academies Press. https://doi.org/10.17226/11338.

NAE. 2023. *Members homepage.* https://www.nae.edu/19581/MembersSection (accessed February 29, 2024).

NASEM (National Academies of Sciences, Engineering, and Medicine). 2012. *Continuing innovation in information technology.* Washington, DC: The National Academies Press. https://doi.org/10.17226/13427.

NASEM. 2016. *Continuing innovation in information technology: Workshop report.* Washington, DC: The National Academies Press. https://doi.org/10.17226/23393.

NASEM. 2020. *Information technology innovation: Resurgence, confluence, and continuing impact.* Washington, DC: The National Academies Press. https://doi.org/10.17226/25961.

NASEM. 2023. *Extraordinary engineering impacts on society: Proceedings of a symposium.* Washington, DC: The National Academies Press. https://doi.org/10.17226/26847.

Nelson, A. 2023. *3D-food printing for healthy eating and delicious desserts.* https://new.nsf.gov/science-matters/3d-food-printing-healthy-eating-delicious-desserts (accessed February 28, 2024).

Nelson, L. K., and K. Zippel. 2021. From theory to practice and back: How the concept of implicit bias was implemented in academe, and what this means for gender theories of organizational change. *Gender and Society* 35(3):330–357.

NIBIB (National Institute of Biomedical Imaging and Bioengineering). 2016. *Rehabilitation engineering*. National Institutes of Health. https://www.nibib.nih.gov/science-education/science-topics/rehabilitation-engineering (accessed April 16, 2024).

NIH. 2011. *NOT-EB-11-006: Notice of NIH Participation in the National Robotics Initiative (NRI)*. https://grants.nih.gov/grants/guide/notice-files/NOT-EB-11-006.html (accessed June 5, 2024).

NIHF (National Inventors Hall of Fame). 2014. *NIHF Inductee Mildred Dresselhaus Invented the Lattice Structure*. https://www.invent.org/inductees/mildred-dresselhaus (accessed May 30, 2024).

NRC (National Research Council). 1975. *Materials and man's needs: Materials science and engineering—Volume I: The history, scope, and nature of materials science and engineering*. Washington, DC: National Academy Press. https://doi.org/10.17226/10436

NRC. 1987. *Directions in engineering research: An assessment of opportunities and needs*. Washington, DC: National Academy Press. https://doi.org/10.17226/1035 (accessed 02-18-2023).

NRC. 1995. *Evolving the High Performance Computing and Communications Initiative to support the nation's information infrastructure*. Washington, DC: National Academy Press. https://doi.org/10.17226/4948.

NRC. 1999. *Funding a revolution: Government support for computing research*. Washington, DC: National Academy Press. https://nap.nationalacademies.org/download/6323 (accessed February 14, 2023).

NRC. 2002. *Information technology research, innovation, and e-government*. Washington, DC: The National Academies Press. https://doi.org/10.17226/10355.

NRC. 2003. *Innovation in information technology*. Washington, DC: The National Academies Press. https://doi.org/10.17226/10795.

NRC. 2007. *The National Science Foundation's Materials Research Science and Engineering Centers program: Looking back, moving forward*. Washington, DC: The National Academies Press. http://doi.org/ 10.17226/11966.

NRC. 2009. *Assessing the impacts of changes in the information technology R&D ecosystem: Retaining leadership in an increasingly global environment*. Washington, DC: The National Academies Press. https://doi. org/10.17226/12174.

NSF INCLUDES National Network, 2022. *INCLUDES name change honors U.S. Congresswoman Eddie Bernice Johnson*. https://www.includesnetwork.org/blogs/nsf-includes-coordination-hub1/2023/02/07/includes-name-change-eddie-bernice-johnson (accessed April 21, 2024).

NSF MRSEC (NSF Materials Research Science and Engineering Centers). 2017. *Columbia University MRSEC collaboration wins National Geographic award*. https://mrsec.org/news/columbia-university-mrsec-collaboration-wins-national-geographic-award (accessed February 29, 2024).

NSF (National Science Foundation). 1994. *The National Science Foundation: A Brief History*. https://www.nsf.gov/about/history/nsf50/nsf8816.jsp#preface (accessed November 29, 2023).

NSF. 2000a. *Award abstract #8907068: Engineering research center for the Carnegie Mellon Data Storage Systems Center*. https://www.nsf.gov/awardsearch/showAward?AWD_ID=8907068&HistoricalAwards=false (accessed November 8, 2023).

NSF. 2000b. *Multidisciplinary Center for Earthquake Engineering Research (MCEER).* https://www.nsf.gov/pubs/2000/nsf00137/nsf00137m.htm (accessed November 8, 2023).

NSF. 2000c. *Nifty 50.* https://www.nsf.gov/about/history/nifty50/ (accessed February 22, 2024).

NSF. 2000d. *Nifty 50 – The Internet.* https://www.nsf.gov/about/history/nifty50/theinternet.jsp (accessed February 29, 2024).

NSF. 2003. *Award abstract #9731294: Renewal proposal for National Nanofabrication Users Network.* https://www.nsf.gov/awardsearch/showAward?AWD_ID=9731294 (accessed January 24, 2024).

NSF. 2006. *Nanoscale science and engineering.* https://nsf-gov-resources.nsf.gov/about/budget/fy2006/pdf/9-NSF-WideInvestments/36-FY2006.pdf (accessed November 8, 2023).

NSF. 2008. *NSF Launches an ERC for revolutionizing medical implants.* https://www.nsf.gov/news/news_summ.jsp?cntn_id=112180 (accessed February 29, 2024).

NSF. 2010a. *Cyberinfrastructure—A special report.* https://www.nsf.gov/news/special_reports/cyber/Cyberinfrastructure%20_NSF.pdf (accessed February 18, 2023).

NSF. 2010b. *NSF Sensational 60.* https://www.nsf.gov/about/history/sensational60.pdf (accessed February 19, 2024).

NSF. 2013a. *3-D printing and custom manufacturing: from concept to classroom.* https://new.nsf.gov/news/3-d-printing-custom-manufacturing-concept (accessed February 28, 2024).

NSF. 2013b. *IIS: Human-Centered Computing (HCC).* https://new.nsf.gov/funding/opportunities/iis-human-centered-computing-hcc (accessed June 5, 2024).

NSF. 2013c. *NSF 13-543: Smart and Connected Health.* https://new.nsf.gov/funding/opportunities/smart-health-biomedical-research-era-artificial/nsf13-543/solicitation (accessed June 5, 2024).

NSF. 2013d. *The engineering behind additive manufacturing and the 3-D printing revolution.* https://www.nsf.gov/news/engineering-behind-additive-manufacturing-3-d (accessed February 29, 2024).

NSF. 2014. *Award abstract #0313747: Center for Collaborative Adaptive Sensing of the Atmosphere (CASA).* https://www.nsf.gov/awardsearch/showAward?AWD_ID=0313747&HistoricalAwards=false (accessed November 9, 2023).

NSF. 2015a. *Mario Rotea to lead NSF Division of Engineering Education and Centers.* https://www.nsf.gov/news/news_summ.jsp?cntn_id=136191 (accessed February 29, 2024).

NSF. 2015b. *NSF engineering research centers: Creating new knowledge, innovators, and technologies for over 30 years.* https://www.nsf.gov/eng/multimedia/NSF_ERC_30th_Anniversary.pdf (accessed November 9, 2023).

NSF. 2017. *Team for research in ubiquitous secure technology (TRUST).* https://www.nsf.gov/awardsearch/showAward?AWD_ID=0424422#:~:text=The%20Team%20for%20Research%20in%20Ubiquitous%20Security%20Technology%20(TRUST)%20was,security%20as%20it%20affects%20society (accessed May 1, 2024).

NSF. 2020a. *FY 2020 Engineering Research Centers program report.* https://www.nsf.gov/pubs/2022/nsf22104/nsf22104.pdf (accessed November 7, 2023).

NSF. 2020b. *I/UCRC: Wind Energy, Science, Technology, and Research (WindSTAR)*. https://www.nsf.gov/awardsearch/showAward?AWD_ID=1362022&HistoricalAwards=false (accessed April 23, 2024).

NSF. 2020c. *Supercomputing future wind power rise*. https://new.nsf.gov/news/supercomputing-future-wind-power-rise (accessed February 29, 2024).

NSF. 2021a. *20 Years of NSF ADVANCE*. https://www.nsf.gov/pubs/2021/nsf21050/nsf21050.pdf (accessed April 19, 2024).

NSF. 2021b. *Broadening Participation in Engineering (BPE)*. https://new.nsf.gov/funding/opportunities/broadening-participation-engineering-bpe (accessed May 31, 2024).

NSF. 2021c. *NSF 22-514: Broadening Participation in Engineering (BPE): Program Solicitation*. https://new.nsf.gov/funding/opportunities/broadening-participation-engineering-bpe/nsf22-514/solicitation (accessed May 31, 2024).

NSF. 2022a. *2022–2026 strategic plan*. https://www.nsf.gov/pubs/2022/nsf22068/nsf22068.pdf (accessed April 16, 2024).

NSF. 2022b. *Battery recycling breakthrough*. https://new.nsf.gov/science-matters/battery-recycling-breakthrough (accessed February 29, 2024).

NSF. 2022c. *Faculty Early Career Development Program (CAREER)*. https://new.nsf.gov/funding/opportunities/faculty-early-career-development-program-career (accessed May 31, 2024).

NSF. 2022d. NSF accelerates use-inspired solutions for persons with disabilities. The National Science Foundation. https://beta.nsf.gov/news/nsf-accelerates-use-inspired-solutions-persons

NSF. 2022e. *NSF and DOE announce partnership for science and engineering research*. https://www.nsf.gov/news/news_summ.jsp?cntn_id=305100 (accessed February 29, 2024).

NSF. 2022f. *NSF History Wall*. https://www.nsf.gov/about/history/historywall/NSF_history-guide_July2022_Update.pdf (accessed February 19, 2024).

NSF. 2023a. *Artificial intelligence at NSF*. https://www.nsf.gov/cise/ai.jsp (accessed November 7, 2023).

NSF. 2023b. Diversity and STEM: Women, Minorities, and Persons with Disabilities. https://ncses.nsf.gov/pubs/nsf23315/report

NSF. 2023c. *Engineering of Biomedical Systems*. https://new.nsf.gov/funding/opportunities/engineering-biomedical-systems (accessed May 30, 2024).

NSF. 2023d. *Engineering research centers*. https://www.nsf.gov/eng/eec/erc.jsp (accessed November 7, 2023).

NSF. 2023e. *NSF invests $162 million in research centers to accelerate materials science from lab to factory*. https://new.nsf.gov/news/nsf-invests-162-million-research-centers (accessed February 29, 2024).

NSF. 2023f. *Science and technology centers*. https://new.nsf.gov/od/oia/ia/stc#:~:text=In%201989%2C%20the%20National%20Science,changing%20needs%20of%20science%20and (accessed January 25, 2024).

NSF. 2024. *Energy technology*. https://new.nsf.gov/focus-areas/energy (accessed February 29, 2024).

NSF. n.d.-a. *About the IUCRC program*. https://iucrc.nsf.gov/about/ (accessed November 29, 2023).

NSF. n.d.-b. *Center for Wind Energy Science, Technology and Research (WindSTAR).* https://iucrc.nsf.gov/centers/wind-energy-science-technology-and-research/ (accessed February 29, 2024).

NSF. n.d.-c. *NSF Award Search: Advanced Search Results for Awards funded by BPE.* https://www.nsf.gov/awardsearch/advancedSearchResult?ProgEleCode=768000&BooleanElement=Any&BooleanRef=Any&ActiveAwards=true#results (accessed May 31, 2024).

NSF. n.d.-d. *NSF award search: Advanced search results for Karan Watson.* https://www.nsf.gov/awardsearch/advancedSearchResult?PIId=&PIFirstName=karan&PILastName=watson&PIOrganization=&PIState=&PIZip=&PICountry=&ProgOrganization=&ProgEleCode=&BooleanElement=All&ProgRefCode=&BooleanRef=All&Program=&ProgOfficer=&Keyword=&AwardNumberOperator=&AwardAmount=&AwardInstrument=&ActiveAwards=true&ExpiredAwards=true&OriginalAwardDateOperator=&StartDateOperator=&ExpDateOperator= (accessed March 1, 2024).

NSF. n.d.-e. *NSF award search: Advanced search results for Stephen Director.* https://www.nsf.gov/awardsearch/advancedSearchResult?PIId=&PIFirstName=Stephen&PILastName=director&PIOrganization=&PIState=&PIZip=&PICountry=&ProgOrganization=&ProgEleCode=&BooleanElement=All&ProgRefCode=&BooleanRef=All&Program=&ProgOfficer=&Keyword=&AwardNumberOperator=&AwardAmount=&AwardInstrument=&ActiveAwards=true&ExpiredAwards=true&OriginalAwardDateOperator=&StartDateOperator=&ExpDateOperator= (accessed March 1, 2024).

NSF. n.d.-f. NSF award search: Award #9153745—AMP: Texas Alliance for Minority Participation. https://www.nsf.gov/awardsearch/showAward?AWD_ID=9153745&HistoricalAwards=false (accessed March 1, 2024).

NSF. n.d.-g. *NSF's history and impacts: A brief timeline.* https://new.nsf.gov/about/history#:~:text=The%20first%20was%20over%20the,draining%20funds%20from%20basic%20research (accessed January 24, 2024).

NSF. n.d.-h. Rehabilitation research at NSF. https://www.nsf.gov/eng/rehab.jsp (accessed April 16, 2024).

NSF. n.d.-i. *Secure and trustworthy cyberspace (SaTC).* https://new.nsf.gov/funding/opportunities/secure-trustworthy-cyberspace-satc (accessed May 1, 2024).

NSF. n.d-j. *Solar energy: Harnessing the sun's power.* https://www.nsf.gov/news/special_reports/solarscience/energy.jsp (accessed June 3, 2024).

Núñez, A. M., J. Rivera, J. Valdez, and V. B. Olivo. 2021. Centering Hispanic-serving institutions' strategies to develop talent in computing fields. *Tapuya: Latin American Science, Technology and Society* 4(1):1842582.

O'Reagan, D.O., and L. Fleming. 2018. The FinFET breakthrough and networks of innovation in the semiconductor industry, 1980–2005. *Technology and Culture* 59(2):251-288.

OSTP (White House Office of Science and Technology Policy). 2024. *The White House Advances Biotechnology and Biomanufacturing Leadership with the Launch of the National Bioeconomy Board.* https://www.whitehouse.gov/ostp/news-updates/2024/03/22/the-white-house-advances-biotechnology-and-biomanufacturing-leadership-with-the-launch-of-the-national-bioeconomy-board/ (accessed May 30, 2024).

Page-Reeves, J., A. Marin, M. Bleecker, M. L. Moffett, K. DeerInWater, S. EchoHawk, and D. Medin. 2017. From community data to research archive: Partnering to increase and sustain capacity within a native organisation. *Gateways: International Journal of Community Research and Engagement* 10:283–297.

Partnerships for Research and Education in Materials. n.d. *Program overview.* https://prem-dmr.org/program-overview (accessed November 8, 2023).

Peña V, Lal B, Micali M. 2014. U.S. Federal Investment in the Origin and Evolution of Additive Manufacturing: A Case Study of the National Science Foundation. *3D Printing and Additive Manufacturing* 1(4): 185–193. https://doi.org/10.1089/3dp.2014.0019

Perkins, R. 2023. *How a small class at Caltech helped launch a computer revolution.* https://www.caltech.edu/about/news/how-a-small-class-at-caltech-helped-launch-a-computer-revolution (accessed February 29, 2024).

President's Science Advisory Committee. 1958. *Strengthening American science: A report.* Washington, DC: U.S. Government Printing Office.

Preston, L., and C. Lewis. 2020. *Agents of change: NSF's engineering research centers: A history.* https://erc-history.erc-assoc.org/ (accessed November 29, 2023).

Pritchard, H. 2011. *Schools of fish help squeeze more power from wind farms.* BBC News, August 8. https://www.bbc.com/news/science-environment-14452133 (accessed February 29, 2024).

pwc. n.d. *Sizing the prize What's the real value of AI for your business and how can you capitalise?* https://www.pwc.com/gx/en/issues/analytics/assets/pwc-ai-analysis-sizing-the-prize-report.pdf (accessed June 4, 2024).

PyrAmes. 2024. *Our pipeline: Creating a new standard of cNIBP.* https://pyrameshealth.com/our-platforms/our-pipeline/ (accessed March 1, 2024).

Raviv, S. 2020, January 21. The secret history of facial recognition. *Wired*, January 21. https://www.wired.com/story/secret-history-facial-recognition/ (accessed March 1, 2024).

Roco, M. C. 2011. The long view of nanotechnology development: The National Nanotechnology Initiative at 10 years. *Journal of Nanoparticle Research* 13(2):427–445. https://doi.org/10.1007/s11051-010-0192-z.

Roessner, D., R. Carr, I. Feller, M. McGeary, and N. Newman. 1998. *The role of NSF's support of engineering in enabling technological innovation: PHASE II executive summary.* SRI International: Prepared for the NSF. https://www.nsf.gov/pubs/1999/nsf98154/nsf98154.htm (accessed April 23, 2024).

Rosser, S. V., S. Barnard, M. Carnes, and F. Munir. 2019. Athena SWAN and ADVANCE: Effectiveness and lessons learned. *The Lancet* 393(10171):604–608.

Russakovsky, O., J. Deng, H. Su, J. Krause, S. Satheesh, S. Ma, Z. Huang, A. Karpathy, A. Khosla, M. Bernstein, A. C. Berg, and L. Fei-Fei. 2015. ImageNet Large Scale Visual Recognition Challenge. *International Journal of Computer Vision* 115(3): 211–252. https://doi.org/10.1007/s11263-015-0816-y.

SAE Media Group. 2023. *Create the Future design contest: 2014 winners.* https://contest.techbriefs.com/2014/winners (accessed February 28, 2024).

SBIR (Small Business Innovation Research). n.d. *Zyrobotics LLC.* https://www.sbir.gov/sbc/zyrobotics-llc (accessed April 16, 2024).

Seeley, D. 2021. *Eddie Bernice Johnson on the importance of STEM education in underserved communities.* Dallas Innovates. https://dallasinnovates.com/eddie-bernice-johnson-on-the-importance-of-stem-education-in-underserved-communities/ (accessed March 1, 2024).

Sewald, J. 2019. Farnam Jahanian, President of Carnegie Mellon University. *Pittsburgh Quarterly*. https://pittsburghquarterly.com/articles/farnam-jahanian-president-of-carnegie-mellon-university/ (accessed May 29, 2024).

Sheela, U. B., Usha, P. G., Joseph, M. M., Melo, J. S., Thankappan Nair, S. T., and Tripathi, A. 2021. *3D printing in dental implants*. In D. J. Thomas & D. Singh (Eds.), 3D Printing in Medicine and Surgery (pp. 83–104). Woodhead Publishing. https://doi.org/10.1016/B978-0-08-102542-0.00007-5

Shen, K. 2024. *Achievements from the nanoscale to the Institute: Paula Hammond SB '84 PhD '93 named 52nd Killian Award recipient*. https://thetech.com/2024/04/11/hammond-killian-lecture (accessed May 30, 2024).

Siewiorek, D. P. 1991. *Oral history: Daniel Siewiorek* [Telephone]. https://ethw.org/Oral-History:Daniel_Siewiorek (accessed March 1, 2024).

Skalak, R., and C. F. Fox. 1998. *Tissue engineering: Proceedings of a workshop held at Granlibakken, Lake Tahoe, California, February 26-29, 1988*. New York: Academic Press.

Southern California Biomedical Council. 2020. *Yulun Wang, Ph.D. EE '88: SoCalBio*. https://www.ece.ucsb.edu/news/all/2020/yulun-wang-phd-ee-88-socalbio-honor (accessed March 1, 2024).

Spafford, E. H. 2003. A failure to learn from the past. *19th Annual Computer Security Applications Conference Proceedings*. 217–231. https://www.acsac.org/2003/papers/classic-spafford.pdf (accessed April 16, 2024).

Sproull, R. L. 1987. The early history of the materials research laboratories. *Annual Reviews of Materials Science* 17:1–12.

Squishy Robotics. 2024. *Squishy robotics—Life-saving, cost-saving robots*. https://squishy-robotics.com/ (accessed March 1, 2024).

Stewart, A. J., and V. Valian. 2018. *An inclusive academy: Achieving diversity and excellence*. Cambridge, MA: MIT Press.

SURE (Summer Undergraduate Research in Engineering/Sciences). 2024. *SURE program overview*. https://sure.gatech.edu/home (accessed April 21, 2024).

TAMEST (Texas Academy of Medicine, Engineering, Science & Technology). 2013. *Q&A with Dr. Joseph Beaman: 3-D printing pioneer*. https://tamest.org/news/3D-printing-pioneer/ (accessed February 28, 2024).

TAMUS LSAMP. n.d. *Louis Stokes Alliance for Minority Participation*. https://tamuslsamp.org/ (accessed March 1, 2024).

Taraborelli, D. 2013. Researching collaboration for a better world: John T. Riedl (1962–2013). *Diff*, July 18. https://diff.wikimedia.org/2013/07/18/researching-collaboration-better-world-john-riedl-1962-2013/ (accessed March 1, 2024).

TechArchives. 2015. *Dennis Jennings*. https://techarchives.irish/how-the-internet-came-to-ireland-1987-97/dennis-jennings/ (accessed February 20, 2023).

Texas A&M University. n.d. *Karan L. Watson*. https://engineering.tamu.edu/electrical/profiles/kwatson.html (accessed March 1, 2024).

Thiry, H. 2017. The importance of community, belonging and support: Lessons learned from a decade of research on Hispanic retention in STEM. *Computing Alliance of Hispanic Serving Institutions. Draft white paper.*

Thomason, R. 2020. Logic and artificial intelligence. The Stanford Encyclopedia of Philosophy Summer. E. N. Zalta (ed.). https://plato.stanford.edu/archives/sum2020/entries/logic-ai/>. (accessed June 23, 2023).

SRI International. 2008. *National and regional economic impacts of Engineering Research Centers: A pilot study, Summary report.* SRI Project P16906.

UC Davis (University of California at Davis). 2023. *Gary S. May | UC Davis Leadership.* https://leadership.ucdavis.edu/chancellor/gary-s-may (accessed May 29, 2024).

U.S. Electrical Information Administration. 2023. *Wind Explained. History of Wind Power.* https://www.eia.gov/energyexplained/wind/history-of-wind-power.php (accessed February 28, 2024).

UN (United Nations). 2022. *World population prospects 2022.* https://population.un.org/wpp/Download/Standard/MostUsed/ (accessed February 29, 2024).

United States. 1968. *Public Law 90-407.* 82 Stat. 363. https://www.govinfo.gov/content/pkg/STATUTE-82/pdf/STATUTE-82-Pg360.pdf

University of Massachusetts. 2023. *CASA: Engineering Research Center for Collaborative Adaptive Sensing of the Atmosphere.* http://www.casa.umass.edu/ (accessed November 9, 2023).

University of Pittsburgh. n.d.-a. *Human Engineering Research Laboratories.* https://www.herl.pitt.edu/education/undergrad/aspire (accessed April 16, 2024).

University of Pittsburgh. n.d.-b. REU projects. https://www.herl.pitt.edu/education/undergrad/aspire/projects (accessed April 16, 2024).

University of Pittsburgh. n.d. *Rory Cooper | School of Health and Rehabilitation Sciences | University of Pittsburgh School of Health and Rehabilitation Sciences.* https://www.shrs.pitt.edu/people/rory-cooper (accessed May 29, 2024).

UT (University of Texas at Austin). n.d. *Joseph Beaman.* https://www.me.utexas.edu/people/faculty-directory/beaman (accessed February 28, 2024).

UTD (University of Texas at Dallas). n.d.-a. *Department of Mechanical Engineering people: Mario A. Rotea.* https://me.utdallas.edu/people/mario-rotea/ (accessed April 23, 2024).

UTD. n.d.-b. *WindSTAR—UTD Wind.* https://wind.utdallas.edu/research/windstar/ (accessed February 29, 2024).

Vanderbilt University. 2017a. *Bara Cola: From VU walk-on fullback to nationally acclaimed engineer.* https://vucommodores.com/bara-cola-from-vu-walk-on-fullback-to-nationally-acclaimed-engineer/ (accessed February 29, 2024).

Vanderbilt University. 2017b. *NSF recognizes double ME alum with Waterman Award.* https://engineering.vanderbilt.edu/news/2017/nsf-recognizes-me-alum-waterman-award/ (accessed February 29, 2024).

Vantage Market Research. 2024. *Additive Manufacturing Market.* https://www.vantagemarketresearch.com/industry-report/additive-manufacturing-market-2349 (accessed June 3, 2024).

Villa, E. Q., 2017. *CAHSI-INCLUDES relevant scholarship for 2-4 year partnerships panel.* Pp. 62-63 in NSF INCLUDES Conference: Advancing the Collective Impact of Retention and Continuation Strategies for Hispanics and other Underrepresented Minorities in STEM Fields. https://par.nsf.gov/servlets/purl/10049620 (accessed April 21, 2024).

Villa, E. Q., S. Hug, H. Thiry, D. S. Knight, E. F. Hall, and A. Tirres. 2019. *Broadening participation of Hispanics in computing: The CAHSI INCLUDES alliance.* In 2019 CoNECD-The Collaborative Network for Engineering and Computing Diversity. https://peer.asee.org/broadening-participation-of-hispanics-in-computing-the-cahsi-includes-alliance.

Villa, E., A. Gates, S. Kim, and D. Knight. 2020. *The CAHSI INCLUDES alliance: Realizing collective impact*. In Zone 1 Conference of the American Society for Engineering Education.

Viola, J., B. Lal, and O. Grad. 2003. *The emergence of tissue engineering as a research field*. The National Science Foundation. Available at https://nsf.gov/pubs/2004/nsf0450/start.htm (accessed April 16, 2024).

Viola, P., and M. J. Jones. 2004. Robust real-time face detection. *International Journal of Computer Vision* 57(2):137–154. https://doi.org/10.1023/B:VISI.0000013087.49260.fb.

Ware, W. H. 1970. *Security controls for computer systems: Report of the Defense Science Board Task Force in Computer Security*. RAND Corporation. https://apps.dtic.mil/sti/citations/tr/ADA076617 (accessed May 1, 2024).

Ware, W. H., and B. Peters. 1967. Security and privacy in computer systems. *Proceedings of the 1977 Spring Joint Computer Conference*. 30:27–282. https://doi.org/10.1145/1465482.1465523.

Weber C. L., V. Peña, M. K. Micali, E. Yglesias, S. Rood, J. A. Scott, B. Lal. 2013. The role of the National Science Foundation in the Origin and Evolution of Additive Manufacturing in the United States. Institute for Defense Analyses Science & Technology Policy Institute. https://www.ida.org/-/media/feature/publications/t/th/the-role-of-the-national-science-foundation-in-the-origin-and-evolution-of-additive-manufacturing-in/ida-p-5091.ashx (accessed February 28, 2024).

Weber, C., V. Peña, M. Micali, B. Lal, E. Yglesias, S. Rool, and J. Scott. 2013. *U.S. federal investment in the origin and evolution of additive manufacturing: A case study of the National Science Foundation*. Institute for Defense Analyses. https://www.liebertpub.com/doi/10.1089/3dp.2014.0019 (accessed February 7, 2023).

Werewool. 2020. *Executive summary prepared for the Biomimicry Institute*. https://biomimicry.org/wp-content/uploads/2020/05/Werewool.pdf.

White House. 2012. President Obama Honors Outstanding Early-Career Scientists. https://obamawhitehouse.archives.gov/the-press-office/2012/07/23/president-obama-honors-outstanding-early-career-scientists (accessed February 29, 2024).

White House. 2015. *President Obama honors outstanding science, mathematics, and engineering mentors*. March 27. https://obamawhitehouse.archives.gov/the-press-office/2015/03/27/president-obama-honors-outstanding-science-mathematics-and-engineering-m (accessed April 21, 2024).

White House. 2022a. Executive order on advancing biotechnology and biomanufacturing innovation for a sustainable, safe, and secure American bioeconomy. https://www.whitehouse.gov/briefing-room/presidential-actions/2022/09/12/executive-order-on-advancing-biotechnology-and-biomanufacturing-innovation-for-a-sustainable-safe-and-secure-american-bioeconomy/ (accessed April 16, 2024).

White House. 2022b. *Fact Sheet: Biden-Harris Administration driving U.S. battery manufacturing and good-paying jobs*. https://www.whitehouse.gov/briefing-room/statements-releases/2022/10/19/fact-sheet-biden-harris-administration-driving-u-s-battery-manufacturing-and-good-paying-jobs/ (accessed February 29, 2024).

WindSTAR. 2023. *WindSTAR 2022 annual report*. https://www.uml.edu/docs/2022-WindSTAR-Annual-Report_tcm18-363014.pdf (accessed April 23, 2024).

Wohlers, T., and T. Gornet. 2015. *History of additive manufacturing*. Wohlers Associates Inc. https://wohlersassociates.com/wp-content/uploads/2022/08/history2015.pdf (accessed April 17, 2024).

Wondwossen W. 2020. *#NSFstories: Breaking Boundaries and Escaping Boxes.* https://new.nsf.gov/science-matters/nsfstories-breaking-boundaries-escaping-boxes (accessed February 28, 2024).

World Energy Council. 2019. *World energy scenarios 2019: Exploring innovation pathways to 2040.* https://www.worldenergy.org/assets/downloads/2019_Scenarios_Full_Report.pdf?v=1571307 963 (accessed April 23, 2024).

Wyss Institute. 2013. *Printing tiny batteries.* https://wyss.harvard.edu/news/printing-tiny-batteries/ (accessed February 28, 2024).

Yost, J. R. 2015. The origin and early history of the computer security software products industry. *IEEE Annals of the History of Computing* 37(2):46–58.

Yost, J. R. 2016. The march of IDES: A history of intrusion detection expert systems. *IEEE Annals of the History of Computing* 38(4):42–54.

Zehnder, S., and J. Kulwatno. 2022. *Biomedical engineering makes strides in assistive technologies.* National Science Foundation. https://beta.nsf.gov/science-matters/biomedical-engineering-makes-strides-assistive (accessed April 16, 2024).

Zippel, K., and M. M. Ferree. 2019. Organizational interventions and the creation of gendered knowledge: U.S. universities and NSF ADVANCE. *Gender, Work and Organization* 26(6):805–82.

5

Communicating Engineering Impacts on Society
to Diverse Audiences

This last chapter of the report addresses the final element of the statement of task: providing guidance on how to inform diverse audiences about engineering's impacts on society. The text examines the considerations that go into communicating impacts and what research tells us about this topic and enumerates the many audiences there are for such communication. It sets forth the committee's approach to developing example outreach materials, presents summaries of those materials, and closes with the committee's conclusions and recommendations regarding communication issues.

The chapter is not intended to be a literature review or a comprehensive examination and is instead an overview of salient information that factored into the committee's thinking. Appendix B contains a more detailed discussion of the literature related to science and engineering communications as part of the documentation accompanying the examples of engineering impacts on society outreach materials.

THE IMPORTANCE OF COMMUNICATING
ENGINEERING'S IMPACTS ON SOCIETY

Any effort to convey the extraordinary impacts that engineering has had on society encounters an immediate problem: most Americans, including most K–12 students and their teachers, have very little understanding of what engineers do or how their work affects everyday life. A survey conducted in Canada ranked engineering second to last in familiarity among a list of professions, behind nurses, doctors, dentists, electricians, lawyers, and accountants and ahead only of architects (Innovative Research Group, 2017). They were roughly in the middle of these professions in terms of favorability, trust, and respect, though those who were more familiar with engineering gave the profession higher rankings than did other respondents. In this survey, engineers were ranked high in their levels of expertise but low in creating social value or being involved in communities.

Polls conducted in the United States have similarly shown that the public[109] has limited understanding of engineering's role in improving quality of life. In a 2004 survey, engineers were ranked far behind scientists in terms of saving lives, being sensitive to societal concerns, and caring about the community (Baranowski and Delorey, 2007). K–12 students in particular tend to have a relatively narrow understanding of engineering. They may know that engineers

---

[109] While the committee uses the collective term "public" here and elsewhere in the report, it acknowledges that there are many "publics" that have different communications needs and different messaging preferences. These are addressed later in the chapter.

design and build things, but they usually cite the mechanical or structural aspects of engineering and overlook the broad range of activities and domains in which engineers are engaged (NAE, 2008). Even many of the students who plan to become engineers have a limited sense of the tasks engineers that perform or the areas in which they work.

Confusion about the differences between science and engineering contributes to inaccuracies in public perceptions. Many of the great technological accomplishments of the 20th century, such as computers, spacecraft, and health technologies, are largely products of engineering but are often described as scientific advances. To take three well-known examples, the Manhattan Project, the Apollo Project, and the Human Genome Project, each succeeded because of the engineering work that brought them to fruition as well as the scientific advances that made it possible to conceive of these projects.

Similarly, almost every object that we use daily has been created or refined by engineering, though the contributions of engineering are often unknown or underappreciated. Scientific findings may or may not have played a role in an engineer's work, and the engineers who designed or modified an object certainly drew on their scientific understanding of the world in carrying out their tasks. But engineering delivered that object into our hands.

The lack of understanding or appreciation for engineering has several negative consequences. Underestimating the value that engineering provides can lead to underfunding of engineering research and education by legislators and government agencies—not only within NSF but also at more mission-oriented federal agencies (the historical underfunding of renewable energy research is an example) and at the state and local level. K–12 students who know relatively little about engineering are unlikely to pursue engineering in college or as a career, even though the profession offers well-paid and fulfilling jobs. Lack of knowledge about engineering can lead to uninformed decisions in the marketplace, which can result in substandard goods and services. In a world facing complex problems that demand creative and technologically sophisticated responses, engineering must be at the forefront of developing solutions.

Importantly, though, these observations do not mean that simply supplying audiences with information on engineering, engineers, or the impacts of engineering on society will address the underlying problems identified. The National Academies report *Communicating Science Effectively: A Research Agenda* (NASEM, 2017) offered insights on the weaknesses inherent in this viewpoint, known as the "deficit model". It observed that effective communication must convey complexity and nuance in a way that is useful and understandable to its intended audience, acknowledging that facts can be interpreted in multiple ways. Technical communication often involves intermediaries like organizations and media, who influence how the information is received based on their credibility, the audience's existing knowledge, and their beliefs. Therefore, simply providing more or clearer information is not always sufficient to achieve communication goals.

Further, assuming that better-crafted messages or more information alone will lead to desired actions overlooks the role of values, socioeconomic conditions, and other considerations in decision-making. People do not base decisions solely on the information presented; thus, effective communication needs to help audiences understand relevant facts and their implications while recognizing the influence of other factors. A well-crafted message for one audience may not suit another, as effective communication requires engaging with different audiences in varying contexts and considering their specific knowledge, needs, and beliefs.

There are myriad impacts of NSF-funded engineering research and education that the public does not associate directly with engineering. While the committee acknowledges that raising awareness of these impacts will not necessarily prompt interest in engineering nor resolve all systemic barriers to higher education access to the field, it may, in some cases, inspire and motivate students to consider an engineering career path by providing a better understanding of engineering and showcasing role models from similar backgrounds. The following sections illustrate how narratives that, by definition, go beyond just factual information, can help to meaningfully connect decision-relevant pieces of information to core audiences to maximize engagement.

CONSIDERATIONS IN COMMUNICATING ENGINEERING IMPACTS

Research on communicating the impacts of engineering to diverse audiences is relatively scarce, but the work that has been done provided insights to the committee to use in formulating their advice to NSF.

Research Conducted for *Changing the Conversation*

In 2008, the National Academy of Engineering (NAE) publication *Changing the Conversation: Messages for Improving Public Understanding of Engineering*, known as the CTC report, presented the results of a research-based effort to develop new and more effective ways of communicating about engineering (NAE, 2008). Using qualitative and quantitative research conducted by the communications and market research firms BBMG, Global Strategy Group, and Harris Interactive, the study began by developing a positioning statement to guide future outreach activities by the engineering community:

> No profession unleashes the spirit of innovation like engineering. From research to real-world applications, engineers constantly discover how to improve our lives by creating bold new solutions that connect science to life in unexpected, forward-thinking ways. Few professions turn so many ideas into so many realities. Few have such a direct and positive effect on people's everyday lives. We are counting on engineers and their imaginations to help us meet the needs of the 21st century (NAE, 2008; p. 5).

This positioning statement was considered to be too complicated and lengthy to serve as the basis for a major communications campaign. Instead, the statement was used to develop a small number of messages designed to present an effective case for the importance of engineering and the value of an engineering education, and these messages were then tested and refined through focus groups and surveys. The four messages that emerged from this process were:

- *Engineers make a world of difference.* From new farming equipment and safer drinking water to electric cars and faster microchips, engineers use their knowledge to improve people's lives in meaningful ways.
- *Engineers are creative problem-solvers.* They have a vision for how something should work and are dedicated to making it better, faster, or more efficient.
- *Engineers help shape the future.* They use the latest science, tools, and technology to bring ideas to life.

- *Engineering is essential to our health, happiness, and safety.* From the grandest skyscrapers to microscopic medical devices, it is impossible to imagine life without engineering.

These messages cast engineering as inherently creative and concerned with human welfare as well as an emotionally satisfying career. Adults and teens of both genders rated "Engineers make a world of difference" as the most appealing message. Among teens, boys found this message as appealing as "Engineers are creative problem solvers." For girls, the second-favorite message was "Engineering is essential to our health, happiness, and safety." Girls ages 16 to 17 in the African American sample and all girls in the Hispanic sample found this second message significantly more appealing than did the boys in those groups.

A fifth message was also tested:

- *Engineers connect science to the real world.* They collaborate with scientists and other specialists (such as animators, architects, or chemists) to turn bold new ideas into reality.

This message was given the fewest votes for "very appealing" among all groups and was the least "personally relevant" for all groups except for African American adults. Messaging about engineering has often emphasized the strong connections between engineering and the need for mathematical and science skills. But by downplaying such aspects of engineering as creativity, teamwork, and communication, messages solely about the scientific and mathematical basis of engineering can imply that it is suited just for a small subset of people with particular interests and characteristics.

The study also developed seven "taglines," short phrases that capture some aspect of the positioning statement in a succinct and memorable way:

1. Turning ideas into reality
2. Because dreams need doing
3. Designed to work wonders
4. Life takes engineering
5. The power to do
6. Bolder by design
7. Behind the next big thing

When survey respondents were asked to rank these taglines, the most successful among all groups was "Turning dreams into reality," though a greater percentage of boys than girls favored it. "Because dreams need doing" was especially popular among teenagers and was liked equally well by girls and boys. Among African American teens, the second most appealing tagline was "Designed to work wonders." The second favorite choice of Hispanic adults and teens was "Because dreams need doing."

The NAE committee that produced *Changing the Conversation* recommended that the engineering community use the positioning statement, messages, and taglines in the report to carry out coordinated and consistent communication to a variety of audiences, including school children, their parents, teachers, and counselors, about the role, importance, and career potential of engineering. It also observed that the messages and taglines should be embedded within a larger strategic framework—a communications campaign driven by a strong brand position communicated in a variety of ways, delivered by a variety of messengers, and supported by dedicated resources. The committee recommended strategies and tools that the engineering community could use to conduct more effective outreach. In addition, it noted that the impacts of communications should be measured so that outreach efforts can be continually improved.

Building on *Changing the Conversation*

The 2013 NAE report *Messaging for Engineering: From Research to Action* reviewed the progress that had been made in implementing the CTC messages and taglines and provided an action plan for each major stakeholder in the engineering community (NAE, 2013). It noted that many organizations had either directly used or adapted the messages and related taglines from *Changing the Conversation*. For example, the Society of Women Engineers reworked all its print and web-based messaging to align with the CTC recommendations, and the Engineer Your Life website, which sought to encourage college-bound girls to pursue engineering, used the CTC and similar messages. The NAE linked the CTC messages and taglines to its Grand Challenges for Engineering and developed tools to help disseminate the CTC messages. Large and well-known companies such as DuPont, Exxon Mobil, and Lockheed Martin generated advertising or recruiting materials that reflected the CTC messages. While many of these efforts are no longer active, they succeeded in spreading the CTC messages and taglines widely within the engineering community (Vest, 2011).

In its proposed action plan, *Messaging for Engineering* outlined basic steps that all segments of the engineering community could take to change the conversation around engineering as well as specific steps for individual segments of the community (NAE, 2013). For example, it urged that communications make explicit use of the words "engineer" and "engineering" and use the CTC messages and taglines with a view of the engineering profession as a whole. It also noted the importance of reaching out to girls, African Americans, Hispanics, and American Indians, given their continued underrepresentation in the field. It urged that the impact of using or adapting the CTC messages be assessed in terms of not just inputs, such as the number of visits to a website, but also outputs, including changes in attitudes or behavior that result from the messaging.

In a paper that came out the same year, Betty Shanahan, who was a member of the committee that produced *Messaging for Engineering*, and two colleagues urged for engineers to use methods and tools for outreach projects that they routinely use in their technical projects but do not typically apply to outreach projects, e.g., research, training, adoption of best practices, and awareness of user needs and culture (Bogue et al., 2013). They observed that engineering professional societies invest an estimated $400 million annually in outreach messaging, activities, and materials to increase the number of students who study engineering in college, particularly among women and members of minority groups. But "this impressive outlay of resources generates a dismal return when one considers the primary outcomes metric: how many students chose to enter, and then persist in, engineering" (p. 12). Reports on which kinds of activities and practices work best, research into messaging and its effects among different groups, and meaningful evaluation of assessment results can all improve the return on investment, they wrote. "Assessment-based practice is the core of successful event planning and delivery in outreach, just as it is in every engineer's career success" (p. 15).

Branding

Both *Changing the Conversation* and *Messaging for Engineering* explicitly sought to change the branding of engineering (NAE, 2008, 2013). As the two reports explain, a brand is an association of specific traits in a person's mind that induces particular ideas and behaviors. Changing a brand therefore can change how a person thinks about and responds to information about a subject.

In a 2011 article in *The Bridge*, Mitch Baranowski discussed some of the issues associated with branding in the context specifically of engineering (Baranowski, 2011). Audiences are co-owners of brands, he pointed out, as they bring their own pre-existing frames to communications. These frames are built not just on fact but also on narratives. According to Baranowski, "Humans are emotional creatures who don't necessarily respond to facts. We respond to stories. We respond to seeing people like us in situations we want to be in. We are aspirational, and this must be taken into account in creative communications. The best brands close the gap between the strategic and the creative, between the rational and irrational" (p. 15).

Brands are built from the inside out, Baranowski observed. Thus, a brand needs to be sold internally before it can be effectively sold externally. Brands should also be strategic drivers, he said, so that the idea behind a brand drives not just current actions based on past planning but ongoing planning for future actions.

Brands need to work across all media types. "From a design perspective, this means brand unity, consistency, developing and following brand guidelines, and internally monitoring how well the guidelines are followed," Baranowski wrote (Baranowski, 2011, p.14). Also, brand ambassadors can be particularly effective in representing a brand to a target audience, but this representation needs to be coherent and consistent. The engineering community is decentralized and varied, encompassing engineering schools, professional societies, companies, the public sector, science and technology centers, and some K–12 schools and programs. Each of these parts of the community has unique goals, capabilities, and parochial interests. In addition, engineers do not agree on a single definition of engineering, or they may focus on their own field rather than engineering in general. These factors create obvious challenges to the coordinated and consistent delivery of messages.

The messages of a branding campaign need to align with the experiences of those at whom the campaign is directed, Baranowski said. For example, undergraduate engineering programs do not necessarily align with the messages of the CTC campaign. An even greater risk is known as the "promise gap," or the "bait-and-switch" (Lachney and Nieusma, 2015), where a communications campaign promises something that subsequent experiences do not deliver. If students decide to study engineering because they have heard that it is a creative profession in which they can solve societal problems but then undergo years of classes in which they mostly work on problem sets, they may feel that they have been misled and could convey that disillusionment to others.

In the context of engineering, Baranowski asked several pertinent questions. To the extent that a brand is associated with a single word, which word does engineering own? What one thing sets engineering apart from other professions? In what way is engineering in a leadership position among professions? How can stories be built around the brand of engineering? Who are the best messengers or ambassadors for engineering? What images should they use, and how should their messages differ for different audience?

Framing

Framing is defined as "casting information in a certain light to influence what people think, believe, or do" (NASEM, 2017; p. 36). When we are provided with information, we use our cognitive schema[110] to assess it and act on it. Information that is consistent with our schema tends to be accepted and assimilated, which in turn reinforces it. Information that does not fit may be downplayed or ignored.

Framing is a process that communicators use to position and present information in a way that takes account of how people interpret it. For instance, a writer scripting a movie might present an engineer as intelligent, industrious, socially disengaged, or beholden to special interests, depending on the frame the writer wishes to invoke. Changing widespread impressions and ideas about engineering therefore requires framing communications in such a way that existing frames are modified to become more realistic and more constructive.

Framing can reorient a person's thinking about an issue, but framing that seeks to change ideas and behaviors must take into account their existing ways of processing information. This requires identifying and analyzing those ways of thinking, including how they change over time (Chong and Druckman, 2007).

Scheufele (2014) summarizes the issues and goals surrounding framing in science communication:

> The challenge for science communication, therefore, is not to debate whether we should find better frames with which we can present science to the public …. Instead, we should focus on what types of frames allow us to present science in a way that opens two-way communication channels with audiences that science typically does not connect with, by offering presentations of science in mediated and online settings that resonate with their existing cognitive schemas, and present issues in a way that "resonates" and therefore is accessible to different groups of nonscientific audiences, regardless of their prior scientific training or interest. (p. 13589)

Storytelling

As Baranowski observed in his analysis of branding, people tend to respond to narratives more strongly than they respond to facts, an observation strongly supported by research (Baranowski, 2011). For nonexpert audiences, narratives are easier to comprehend than are traditional scientific explanations. Stories feature a cause-and-effect relationship among events that occur to characters over a period of time. In this way, narratives describe specific situations within the context of more general truths. Explanations in science and engineering, in contrast, tend to move from abstract truths toward specific instances that adhere to those truths. As Dahlstrom observes, "[i]n essence, the utilization of logical-scientific information[111] follows deductive reasoning, whereas the utilization of narrative information follows inductive reasoning" (Dahlstrom, 2014; p. 13614).

Research has shown that narratives are easier to understand and remember compared with standard technical and scientific communication (Schank and Abelson, 1995). Our propensity to rely on narratives may be a product of evolution, where narratives conferred an adaptive

---

[110] *Cognitive schema* are mental structures or models we develop over time from our experiences and knowledge. These schemas help us organize and interpret all the information we encounter, forming the basis of how we understand the world.

[111] "Logical-scientific information" is defined in the article as information based on facts, while narrative information relies on story-telling.

advantage by enabling humans to construct possible realities, model cause-and-effect relationships, and understand complex social interactions. The use of narratives has been increasing in science education, where they are better able to evoke emotional and personalized responses than conventional instruction (Jones and Crow, 2017). They are also effective in health communications, where people can more easily relate to narratives in understanding and adopting healthier behaviors.

Once people are done with formal schooling, they get most of their information about science and engineering from the mass media. To compete for attention, the mass media tend to rely on stories, anecdotes, personalization, and other narrative structures, reflecting the power of narratives to capture our attention and shape our beliefs.

Engineering is particularly amenable to a narrative approach. Individuals and teams are engaged not only in understanding causes and effects but also in adapting those causes and effects to human advantage. Stories of exploration, effort, enlightenment, and triumph are commonplace in engineering. Such narratives offer many opportunities to convey to non-engineers the extraordinary impacts of engineering on society.

The narrative approach, though, must be used thoughtfully and with full appreciation of its potential pitfalls. Dahlstrom and Ho (2012) identify three primary ethical issues related to the use of narrative in science communication. First, they emphasize the need to consider the underlying purpose of using narrative: whether it is intended for comprehension or persuasion. They highlight that while narratives can make scientific information more relatable and understandable, they can also be used to subtly influence or persuade audiences, which raises ethical concerns about manipulation and the communicator's intent.

The authors also discuss the importance of maintaining accuracy within narratives. They point out that narratives can sometimes oversimplify complex concepts or lead to overgeneralization, which can result in misinformation. Ethical communication should strive to balance the engaging nature of narratives with the necessity of accurate and representative information.

And last, Dahlstrom and Ho pose the question of whether narratives should be used in science communication at all, given their potential to manipulate. They argue that while narratives are a powerful tool for engagement and understanding, their use must be carefully considered to avoid misleading audiences. Science communicators need to weigh the benefits of narrative storytelling against the ethical implications of influencing public perception and behavior through emotionally compelling stories.

Using narratives to communicate scientific and technological information thus requires careful consideration of its ethical dimensions. These considerations include the honesty of the content, where communicators must avoid misrepresentation and provide sufficient context to prevent misleading the audience. Clarity and accessibility are also crucial. Communicators should use understandable language, avoid jargon, and transparently convey the information's limitations and uncertainties. To maintain objectivity, they should be mindful of personal and institutional biases, present balanced views, and disclose any conflicts of interest.

Respect for the audience is paramount, recognizing diverse backgrounds and perspectives, and engaging them in ways that empower critical thinking and informed decision-making. Ethical use of emotional appeal involves balancing emotion with facts and avoiding fearmongering. Cultural sensitivity requires respecting cultural differences and including diverse voices in narratives. Privacy and consent must be respected, particularly in sharing stories involving individuals. Finally, ethical storytelling must consider its impact on the people and communities towards whom the outreach is directed. Addressing these ethical considerations ensures responsible, trustworthy, and respectful communication, ultimately fostering a more informed and engaged public.

REACHING DIVERSE AUDIENCES

Changing the Conversation and *Messaging for Engineering* point out that messages about engineering, to be effective, must be tailored to appeal to their target audience (NAE, 208, 2013). These audiences are briefly noted below, along with groups that have the potential to be communicators.

K–12 Students

While a small but growing number of K–12 students have received formal education in engineering, many others develop conceptions of engineering other ways, including, in part, via the science and mathematics classes they take in middle and high school. This direct and indirect exposure to engineering has the potential to increase awareness of the work of engineers, boost youth interest in pursuing engineering as a career, and increase the technological literacy of all students (NAE/NRC, 2009).

Because it is so well suited to hands-on activities, engineering education can be inherently attractive to K–12 students. It can teach students that problems have many potential solutions, and that careful analysis and modeling can both suggest and differentiate among those solutions. It can complement the teaching of science, mathematics, and other subjects and demonstrate the social, economic, and environmental impacts of technical endeavors. And it can inculcate habits of mind that are essential in the 21st century: systems thinking, creativity, optimism, collaboration, communication, and consideration of ethical issues. Both educators and guidance counselors should be part of this effort.

As mentioned earlier, effective messages about engineering may differ between boys and girls. For example, research suggests that, from an early age, girls are less interested in subjects that are characterized by male gender stereotypes, suggesting the need to address these stereotypes early in children's education (Master et al., 2021).

One existing messaging effort is "Engineers Week", which "is dedicated to ensuring a diverse and well-educated future engineering workforce" through programs intended to stimulate interest in the field, particularly in in K–12 students (NSPE, 2024). DiscoverE, one of the organizations promoting "EWeek", posts educator and volunteer toolkits, event primers, promotional materials, and other resources for download in association with the event (DiscoverE, 2024).

There are also other strategies for engaging young people from all segments of society in the pursuit of engineering as a career, including programs such as Project Lead the Way,[112] which gives grants to elementary, middle, and high schools to implement engineering and other STEM curriculum as well as providing professional development for teachers. These strategies are executed at various points in the academic path and with varying levels of success and access to the necessary resources (people, expertise, and funding) to execute. Some of the most impactful strategies include participation in hands-on, project-based engineering camps, courses,

---

[112] https://www.pltw.org/.

workshops, and work experiences and in some instances represent partnerships between schools, universities, communities, the private sector, and government agencies (American Progress, 2024; NASEM, 2019). The Institute for Broadening Participation hosts on the web a searchable database with more than 1,000 fully funded STEM programs for K–12, undergraduates, graduate students, postdocs, and faculty.[113]

Members of Underrepresented Groups

Women, African Americans, Hispanics, American Indians, people with disabilities, and members of other minority groups remain significantly underrepresented in engineering in the United States. According to the most recent data collected by the National Science Foundation, women still make up only about 16 percent of college-educated engineers in the United States, and in 2020 women, Black or African American, Hispanic or Latino, and American Indian or Alaska Native students received just 24 percent, 5 percent, 14, and 0.3 percent of engineering bachelor's degrees, respectively, well below the representation of these groups among the U.S. college-age population (NCSES, 2023). Some signs of diversification have appeared in recent years; for example, the number of bachelor's degrees awarded to women in engineering more than doubled from nearly 15,000 in 2011 to more than 31,000 in 2020, and from 2017 to 2021 the number of Hispanic students pursuing graduate degrees in science, technology, engineering, and mathematics (STEM) rose by about 50 percent. But much more needs to be done before these and other groups will be represented proportionately in engineering.

Messages about engineering need to be appealing and believable (DiscoverE, 2023) and to excite the interest of students from underrepresented groups if they are to positively affect perceptions, persistence, and the pursuit of engineering careers (NAS/NAE/IOM, 2011). Thus, targeted messages about engineering to underrepresented groups might emphasize the opportunity to make a difference in the world or the chance for everyone to engage in meaningful, collaborative, and creative work, while also providing financial security. Beyond persuasive messages, underrepresented groups also need welcoming and inclusive academic and professional environments to make the make the message that "engineering is for everyone" credible (DiscoverE, 2023). This requires effort across all communities involved in engineering and engineering education.

It is also important to note that messages about engineering targeted to underrepresented groups are more effective when they are delivered by people who belong to the same underrepresented group as the intended audience (DePass and Chubin, 2015). Students more readily empathize with teachers, counselors, mentors, and advisors who are like themselves. Furthermore, representation must go beyond those who stand in front of the classroom. As Shanahan and her colleagues wrote in their 2013 article, "To be successful, organizations must go beyond asking nontraditional members to volunteer; such members must be recruited to leadership positions, have the ability to drive changes in existing programming and the organization overall, have the expectation of being listened to, and be recognized for their contributions to the society and the discipline" (Bogue, et al., 2013, p. 14).

As the U.S. population becomes increasingly diverse, engineering needs to account for the values and concerns of all people. This objective can be achieved most effectively by having an engineering workforce that reflects that diversity, which means reaching the members of

---

[113] https://www.pathwaystoscience.org/.

underrepresented groups with the messages, information, and assistance they need to help them join the field.

Engineering Schools

Engineering schools have lines of communication not only to their own faculty members and students but to K–12 teachers, counselors, and students and to their alumni. There are more than 300 ABET[114]-accredited engineering schools in the United States, all of which perform outreach of some sort to prospective students and the wider world. In addition, these schools have connections to their local communities, to other disciplines within their college, and to national and international engineering organizations. Engineering schools have many ways of tailoring and disseminating messages about the impacts of engineering both on individuals and on society as a whole.

As recommended in *Messaging for Engineering*, engineering schools can hold new-faculty orientation workshops or other faculty training sessions (NAE, 2013). They can direct messages about engineering not only to their own students but to university students who have not decided on a major or are majoring in different subjects. They can work with schools of education so that future K–12 teachers are aware of what engineering is and what engineers do. They can encourage engineering undergraduates to volunteer in K–12 classrooms and get involved in the Grand Challenges for Engineering. Such students can publicly disseminate the message of engineering's impact both during their school years and subsequently.

Science and Technology Centers

Along with formal schooling and the mass media, the hundreds of museums and other science and technology centers in the United States are a major source of information about STEM for children, adolescents, and adults. These organizations have the potential to teach visitors of all ages about engineering and its role in shaping modern life. When designing new exhibits or revising existing ones, they can highlight the role of engineering in engaging and memorable ways. They can also involve engineers from academia, professional societies, and industry in the review of exhibits and in outreach activities. In addition, they have great expertise in messaging about science and technology and could advise the other sectors of the engineering community in how to craft messages that have long-lasting and beneficial effects.

Industry

Technology-related companies rely heavily on engineers, not only for traditional engineering tasks but for company leadership and guidance. Many of these companies undertake education, job recruitment, and marketing outreach efforts to attract skilled personnel and to build the pipeline that will create future engineers. As a result, they have many opportunities to shape the messages and information that people receive about engineering.

Companies can feature engineers and engineering in outreach to K–12 students, colleges, and engineering societies. They can partner with other segments of the engineering community,

---

[114] ABET, formerly the Accreditation Board for Engineering and Technology, is a nonprofit, nongovernmental organization that accredits "college and university programs in the disciplines of applied and natural science, computing, engineering and engineering technology at the associate, bachelor's and master's degree levels" (https://www.abet.org/about-abet/).

such as professional societies and colleges of engineering. They can collaborate among themselves through such organizations as the Business Roundtable or the Council on Competitiveness as well as providing their engineers with the opportunities and support to engage with other parts of the engineering community, including students.

Government Agencies

As with companies, government agencies rely on engineers in carrying out their missions and have many opportunities to shape the messages that students and members of the public receive about engineering. *Messaging for Engineering* urged government agencies to "incorporate the CTC messages in education and outreach programs . . . and in all STEM-related government programs that support hands-on experiments and engineering design activities for schools, libraries, scout troops, civic centers, and other organizations" (NAE, 2013, p. 55). Government agencies can also work with other segments of the engineering community, including industry partners, in outreach programs, and they can create incentives for recipients of federal grants and contracts to convey accurate information about engineering. Engineers employed by the federal government can speak and provide mentoring to students and their teachers. In these and other ways, government agencies can help demonstrate the extraordinary impacts of engineering in the past and the potential for engineering to further improve human life.

Engineering Professional Societies

Hundreds of thousands of engineers belong to professional societies, providing a unique opportunity to coordinate educational and public outreach about engineering. With training, the members of these societies can become effective ambassadors in conveying messages about the value and importance of engineering.

As is recommended in *Messaging for Engineering*, societies should educate their members about public communication and effective messages (NAE, 2013). This can be done through sessions at conferences, through society publications, and through local chapters. Societies can conduct meetings and workshops with policymakers, educators, and members of the public. Messages about the engineering profession as a whole are preferable, but messages about individual disciplines can be helpful as well.

The American Society for Engineering Education has a unique position in representing the profession as a whole rather than specific disciplines. In addition, diversity-based societies provide an opportunity to reach out specifically to groups underrepresented in engineering, with coordination between discipline-based and diversity-based societies enabling greater coordination and dissemination of messages. Box 5-1 provides a list of such organizations.

BOX 5-1
Diversity-Oriented Engineering Organizations

Women-Oriented
American Association of University Women
Society for Women Engineers
Women in Global Science and Technology
Women in Technology International

Racial/Ethnic Minority-Oriented
American Indian Science and Engineering Society (AISES)
Great Minds in STEM (GMiS)
Inclusive STEMM Ecosystems for Equity & Diversity (ISEED)
INROADS
National Action Council for Minorities in Engineering (NACME)
National Association of Multicultural Engineering Program Advocates
National GEM Consortium
National Society of Black Engineers
Society for Advancement of Chicanos/Hispanics and Native Americans in Science (SACNAS)
Society of Asian Scientists and Engineers (SASE)
Society of Hispanic Professional Engineers
Society of Mexican American Engineers and Scientists (Latinos in Science and Engineering)

LGBTQ+-Oriented
500 Queer Scientists
Out in STEM (OSTEM)
Out to Innovate (formerly National Organization of Gay and Lesbian Scientists and Technical
    Professionals [NOGLSTP])
Pride in STEM

Disability-Oriented
AccessSTEM
Entry Point!

THE COMMITTEE'S APPROACH TO
DEVELOPING EXAMPLE OUTREACH MATERIALS

The statement of task directed the committee to "[p]rovide guidance on how to reach and engage diverse audiences . . ., promote better understanding of the vital role of engineering in government, business, and society; and engage young people from all segments of society to encourage pursuing a career in engineering." As noted in Chapter 1, NSF provided additional guidance on this element of the task in their charge to the committee. Dr. Susan Margulies, NSF's Assistant Director for Engineering, indicated that the committee should focus on "stories and people" in its impact descriptions, noting that stories can illustrate how fundamental research and innovative education modalities translate into societal benefits; propel prospective visions and new research directions; embolden creativity and risk; and, most importantly, inspire, motivate, and connect (Margulies, 2021). In keeping with the literature on communications summarized in this chapter, the committee sought to identify messages with this focus that were accurate, accessible, clear, engaging, and relevant to diverse audiences.

Several previous National Academies reports and other scholarship have addressed the challenges of and strategies for communicating scientific and technical information to these audiences, including NAE reports that offered observations specific to engineering. The committee did not repeat this work but rather concentrated on developing information and materials directly relevant to NSF's role in bringing about engineering impacts on society. It was informed by the scholarship summarized in this chapter and the additional outreach-specific literature presented in the Appendix B sections titled "Rationale Behind Content Choices".

In addition to the descriptions of the achievements, stories, and people presented in the impact descriptions presented in Chapter 4, the committee contracted with the Alan Alda Center for Communicating Science (Alda Center) and with the MIT Office of Engineering Outreach Programs to develop a set of example materials for NSF to consider in their outreach efforts. These materials, which were prepared under the supervision of committee members Laura Lindenfeld and Eboney Hearn, address topics and people identified by both the exemplary impacts narratives and in the symposium in order to give NSF a broader range of alternatives with which to work. Note that these materials are presented as examples only and have not been subject to the refinement that would accompany a professional product. Furthermore, they have not been subject to review by the NSF and should not be considered as representing the agency's point of view.

As part of this task, an informal, small-scale assessment of draft materials was conducted with seventh- through twelfth-grade students. These students attended public schools in either Boston, Cambridge, or Lawrence, Massachusetts, and were from underrepresented or underserved backgrounds in STEM. They were participants in a STEM outreach program that gave them exposure to a variety of disciplines in engineering and thus had already shown an interest in exploring topics in engineering and science.

Participants provided feedback via an anonymous questionaire after reviewing the draft materials along with documents that described the content and objectives of the materials. This instrument was composed of the questions below and was administered separately for each example reviewed.

- What did you appreciate about content in sample X?
- What did you wonder about content in sample X? What would you change?
- How was sample X successful in meeting its objective?
- In what ways is media/communication in sample X a good fit for a middle or high school audience? In what ways could it fit better?
- How do you typically engage STEM content inside and outside of school?
- When engaging with STEM content, what type of information have you found to be helpful in building interest and knowledge in STEM?

Responses were taken into consideration by the Alda Center as they refined the outreach materials.

Notable takeaways from the responses included the value of providing videos and blogs with engineers from diverse backgrounds to promote a sense of belonging in engineering and a sense of resilience. Content creators were encouraged to consider using short videos that embedded humor to keep the attention of high schoolers and encourage them to reshare content.

In addition, it was suggested that content be accompanied with resources and information on how to get involved in engineering or STEM programs in the local community or through a formal program.

EXAMPLE OUTREACH MATERIALS

Five examples of outreach materials were developed. They are titled

1. Meet an Engineer
2. Queen of Carbon Science "Family Tree"
3. Earthquake Shake Table
4. Grand Challenges
5. Extraordinary Impacts of Engineering

The examples are aimed at different segments of the general public, with an emphasis on students and on groups that may otherwise be poorly informed about engineering and engineers. They use a variety of media (short videos, interactive graphics, blog posts, workbooks, etc.) and take different approaches (inspirational, educational, humorous, etc.) to engage their audiences. Their content is briefly described in the sections below and summarized in Table 5-1. Complete descriptions of the examples are contained in Appendix B. The text supporting them details the target audiences, content objectives, media format, pitch outline, script, and the rationale behind the content choices, plus additional information and citations to literature consulted. Online materials accompany the examples.

136

TABLE 5-1 Summary of Example Engineering Impacts Outreach Materials Intended for Diverse, General Population Audiences

Title	Intended audience[s]	Media Format	Content Summary
Meet an Engineer	Young people (high school) in historically marginalized groups in STEM (especially underrepresented gender/racial/ethnic groups)	Video sharing social media platform video (2–5 minutes long)	Interview with a high-profile, NSF-funded engineer from an underrepresented group
Queen of Carbon Science "Family Tree" (inspired by the work of Millie Dresselhaus)	High schoolers (especially young women)	Clickable-interactive image-based web post	Graphic of the many forms of carbon connected in a "family tree"-type diagram illustrating their real-world applications
Earthquake Shake Table	General public (teens and up), especially those who live in high-hazard earthquake areas	Video sharing social media platform short-form video (2 minutes)	Video illustrating how structures are endangered by earthquakes and how shake tables can be used to test designs
Grand Challenges in Engineering	Elementary and middle school students	Fill-in-the-blank style workbook for elementary students; blog posts for middle school students	Workbook with prompts encouraging students to identify a problem and then draw a device that solves the problem; blog posts highlighting stories of inspiring engineers
Extraordinary Impacts of Engineering	Gen Z/younger millennials	Video sharing social media platform short-form video (2 minutes)	Animated video illustrating how the technology of an everyday device like a phone has advanced

Example 1: *Meet an Engineer*

This content is aimed at young high school students in STEM, especially those from historically marginalized backgrounds. The objectives are to redefine the image of engineers, emphasizing their social, creative nature and diverse interests beyond work, aiming to resonate with the audience by highlighting shared identities and interests. This is achieved by showcasing profiles of engineers who resemble them and share their passions, thereby fostering interest in engineering careers. Through short-form video content that can be shared on social media platforms, the intention is to link engineering with awe and curiosity, moving beyond mere math and science proficiency and emphasizing problem-solving and creativity [short-form video link]. An interview would be conducted with a high-profile engineer who has received funding from NSF and belongs to an underrepresented group, inspired by the NSF's Science Happens Here campaign. The pre-designed interview, with a sample outline script and questions, would take place during an activity that interests the engineer and also appeals to the target audience. The editing for this interview would highlight the engineer's personal story while also connecting it to broader themes about the essence of engineering and the traits of curiosity and creativity in individuals.

In the sample outreach material, Gary May, the Chancellor of the University of California, Davis, and a former professor and Dean of Georgia Tech's College of Engineering, was interviewed. In addition to highlighting his professional accolades—especially in increasing opportunities for groups that are historically underrepresented in the engineering field—the video includes more personal information, for instance, on May's path to engineering from a family of non-engineers and his life-long interests in comic books, Star Trek, and "ways to imagine a different world."

The rationale behind this content is to emphasize the scientific and societal benefits of diversity in engineering, citing increased productivity, improved cognitive performance, and innovation gains. However, the engineering workforce fails to reflect the diversity of the U.S. population, hindering innovation. To address this, the content seeks to amplify underrepresented voices and challenge stereotypes by featuring an engineer who shares interests and identities with high school students. Drawing from best practices in science communication and social identity theory, the content aims to foster trust through warmth and authenticity. Additionally, it prioritizes culturally relevant communication to ensure accessibility and resonance with the audience's everyday reality. By showcasing diverse narratives and engaging storytelling inspired by initiatives like *The Story Collider*, the content aims to inspire and empower underrepresented groups to pursue careers in STEM.

A still image from the video interview with Dr. May is shown in Figure 5-1.

FIGURE 5-1 Still from the Example 1: *Meet an Engineer* video interview with Dr. Gary May.

Example 2: *Queen of Carbon Science "Family Tree"*

This content is aimed at high school students, especially women, with the objectives of reframing basic engineering research as having significant real-world impacts, challenging the perception of engineers as solitary geniuses by highlighting stories of individuals who have been influenced by and have influenced many others, and thus portraying engineering as a collaborative and social endeavor. Importantly, there is a focus on promoting gender equity by ensuring that the narratives of engineering history include the contributions of women. The content will be in a clickable interactive image format for web posts [interactive graphic download Windows link and Apple OS link]. The example showcases various forms of carbon in a "family tree" diagram, demonstrating their practical applications. Engineers and scientists associated with these materials are depicted within the tree, emphasizing that engineering is a collaborative endeavor rather than the work of a lone prodigy. Dr. Millie Dresselhaus's significant contributions to engineering, particularly those supported by NSF, are highlighted. The interactive graphic depicts a woman, symbolizing Dr. Millie Dresselhaus, under a tree, writing with a pencil. Clicking on the graphic reveals information about carbon, its uses, and the ways in which it inspired Dresselhaus. The graphic zooms in on the pencil tip to discuss graphite and Dresselhaus's specific work with this material. The tree's roots represent carbon storage, leading to insights on graphite compounds and so on. The graphic integrates Dresselhaus's contributions and mentions her influence on Nobel Prize-winning researchers. It concludes with a summary of Dresselhaus's accomplishments as a woman in engineering. The rationale behind the content choices emphasizes the importance of portraying a diverse history of engineering, challenging gender stereotypes, and expanding the image of engineering beyond traditional narratives. It aims to engage girls by highlighting the biographies of female engineers like Millie Dresselhaus and showcasing engineering as a communal, social process. Visuals play a crucial role in engaging audiences, but it is essential to adhere to best practices in visual design to ensure accessibility and effectiveness.

The opening image of the example's interactive graphic is contained in Figure 5-2.

FIGURE 5-2 The opening image of the Example 2: *Queen of Carbon Science* interactive graphic.

Example 3: *Earthquake Shake Table*

This content targets the general public, particularly teens and older individuals, who live in high hazard earthquake areas of the United States. It aims to highlight impressive yet not-widely-known innovations and societal impacts supported by NSF investments, showcasing the "cool-factor" of the country's largest shake table to inspire awe and appreciation for engineering's role in earthquake risk mitigation. Using the popular "Did you know?" format, these short-form videos are suitable for social media platforms such as TikTok, Instagram reels, and YouTube Shorts. The sample content features a video highlighting the largest three-dimensional shake table in the United States, demonstrating how engineers simulate earthquake conditions to strengthen buildings [video link]. The pitch emphasizes that such work is possible thanks to advances in engineering. Optionally, a follow-up video could depict a presenter constructing a model building and testing it on a do-it-yourself shake table (demonstrated in brief in the example video).

The video takes a humorous approach to engage a general audience that is not already interested in engineering, using research on persuasion to change attitudes. Using platforms like TikTok and concise, short-form, "infotainment"-style narration, the content aims to stimulate and maintain the interest of the target audience. The focus on attention-grabbing content—earthquakes—aims to evoke curiosity while being mindful of avoiding audience burnout.

A still from the video illustrating a simple "homemade" shake table is shown in Figure 5-3.

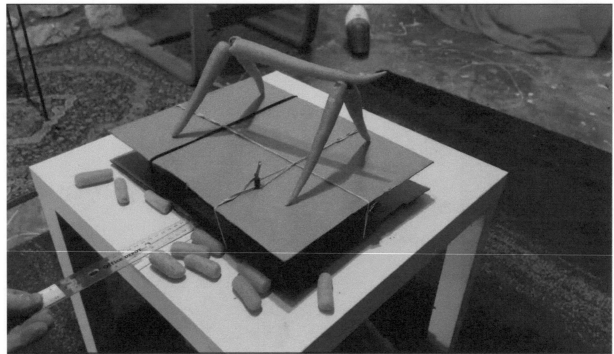

FIGURE 5-3 Still from the Example 3: *Earthquake Shake Table* video.

Example 4: *Grand Challenges in Engineering*

This content is aimed at elementary and middle school students to challenge the misconception that the lack of representation in engineering is solely a "pipeline" issue, emphasizing the need for culturally relevant pedagogy in STEM education. It seeks to inspire young individuals to embrace their cultural identities and bring their whole selves into STEM spaces, fostering inclusivity and diversity within the field. The content would be in two formats, a workbook for elementary school students (not depicted) and blog posts for middle school students [blog post link]. Using the NAE *Grand Challenges for Engineering*[115] as an inspiration, the workbook would aim to demonstrate that anyone can think like an engineer, guiding younger readers through design steps using drawings to solve real-world problems that they care about. The students' drawings might be showcased anonymously on a moderated NSF public website, connecting their contributions to NSF-funded efforts. The workbook would conclude with encouragement to continue applying engineering thinking in their lives. The blog posts would introduce grand challenges facing the world, featuring underrepresented engineers working on solutions and highlighting past and recent innovations. The examples would emphasize the richness of engineering through diversity and the impact of engineers using their backgrounds and perspectives to address global issues. The goal driving the content choices is to engage young people in engineering by using problem-posing methodology, allowing them to take on

---

[115] https://www.engineeringchallenges.org/challenges.aspx.

the role of experts and drawing on their funds of knowledge and cultural connections. Art-based communication enhances creativity and learning outcomes, aligning with multimodal learning approaches. By positioning the audience as holders of knowledge and emphasizing engineering's creative potential, the content is intended to develop participants' engineering identity and increase their self-efficacy in the field. It also highlights the community-minded aspects of engineering to attract historically marginalized students. The content seeks to stimulate interest in engineering by providing both examples and direct experiences, recognizing the importance of embracing diverse forms of knowledge to address society's challenges and empowering young people in engineering spaces.

Two photographs used in the illustrative blog posts are presented in Figure 5-4.

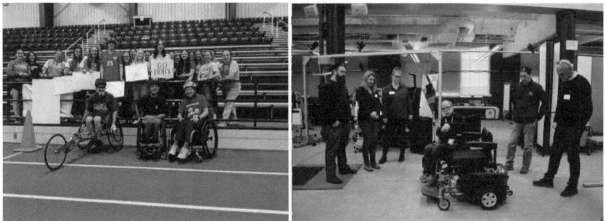

FIGURE 5-4 Photographs from the Example 4: *Grand Challenges in Engineering* blog post. SOURCE: Dr. Rory Cooper.

Example 5: *Extraordinary Impacts of Engineering*

The content is aimed at "Gen Z" and younger millennials with the objective of using narratives to frame engineering innovations as beneficial to people's everyday lives. Using TikTok-style short-form videos/reels [short form video link], the content is intended to engage viewers through nostalgia, humor, and concise language. The introduction sets the scene by showcasing life without such modern technology as smartphones. Given the importance of keeping videos short to maximize engagement, it uses nostalgia, particularly targeting Gen Z and millennials through connections to 2000s "throwbacks," to captivate the audience emotionally and ensure the video's success. Two video outlines are described. The first video presents a journey through time, showcasing clips from different decades depicting people using various technologies, from payphones to smartphones, highlighting the evolution of technology and its impact on our lives. The aim is to illustrate the advances in technology and the role of engineers in shaping our modern world. The second video uses an educational animation in a humorous style, reminiscent of popular educational animations from past decades such as The Disney

Channel's Healthy Handbook, BrainPOP, and Wild Kratts productions. This approach is intended to appeal to nostalgia while providing informative content in an engaging and entertaining format. The content is designed to address the lack of understanding among the general public regarding the role of engineers and the social value of engineering innovations. Drawing from persuasion techniques like emphasis framing, the video highlights the societal impacts of engineering in a humorous manner, encouraging audiences to imagine life without certain technologies. It incorporates participatory elements and interactive platforms such as TikTok to engage audiences effectively, recognizing the importance of co-creating content with users. Additionally, research on humor in science communication informs the use of humor to enhance engagement; that said, caution is advised in order to avoid the video backfiring, especially when correcting misinformation. Ultimately, this content is intended to make engineering more relatable and accessible by framing it as an everyday experience and taking advantage of social media as a popular communication channel for younger audiences.

Figure 5-5 is a still from an example short-form reel style video illustrating how the modern smartphone has evolved to contain a number of functions previously found in separate devices.

FIGURE 5-5 Still from the Example 5: Short-form video illustrating some of the devices contained in the modern smartphone.

CONCLUSIONS AND RECOMMENDATIONS

The committee's statement of task directed it to "offer conclusions and recommendations on how to best promote understanding of engineering's place in society and how NSF contributes to it." The information developed in this chapter and the additional material presented in the "Rationale Behind Content Choices" sections accompanying the examples of engineering impacts on society outreach materials contained in Appendix B led the committee to the following conclusions:

- There is an opportunity to change the way that engineers and engineering are perceived by the general public by highlighting the many ways that engineering affects everyday life, the contributions that engineers make to improving those lives, and the role that investments in engineering education and research play in making these contributions possible.
- Outreach efforts that are grounded in the research on engagement and communication are more likely to reach target audiences and have the intended impact. Research indicates that efforts that use or highlight people with whom or content with which target audiences can relate are more effective.

These materials also form the basis for the following recommendations regarding the communication of engineering impacts on society to diverse audiences:

NSF, in its outreach efforts regarding the support of engineering education and research, should

- **draw upon the literature and experts on public engagement and communication to better target its messaging.**
- **increase the participation and diversity of organizations, people, and voices who have not been well represented in the engineering profession in its messaging.**
- **employ communication forms (short-form videos and media that can be consumed on phones for example) and forums (including social media platforms) that are used by its target audiences.**
- **feature diverse and relatable people and stories that illustrate how engineering is making everyday life better and how engineers improve the lives of others.**
- **incorporate tracking of the effectiveness of specific messaging efforts into outreach efforts.**

REFERENCES

American Progress. 2024. *K-12 work-based learning opportunities: A 50-state scan of 2023 legislative action.* https://www.americanprogress.org/article/k-12-work-based-learning-opportunities-a-50-state-scan-of-2023-legislative-action/ (accessed May 31, 2024).

Baranowski, M. 2011. Rebranding engineering: Challenges and opportunities. *The Bridge* 41(2):12–16. Available online at www.nae.edu/Publications/Bridge/51063/51088.aspx (accessed April 15, 2024).

Baranowski, M., and J. Delorey. 2007. *Because dreams need doing: New messages for enhancing public understanding of engineering.* Project Report for the NAE Committee on Improving Public Understanding of Engineering Messaging.

Bogue, B., E. Cady, and B. Shanahan. 2013. Professional societies making engineering outreach work: Good input results in good output. *Leadership and Management in Engineering* 13:11–26.

Chong, D., and J. N. Druckman. 2007. Framing theory. *Annual Review of Political Science* 10:103–126.

Dahlstrom, M. F. 2014. Using narratives and storytelling to communicate science with nonexpert audiences. *Proceedings of the National Academy of Sciences* 111(Suppl 4):13614–13620.

Dahlstrom, M. F., and S. S. Ho. 2012. Ethical considerations of using narrative to communicate science. *Science Communication 34*(5):592-617.

DiscoverE. 2023. *Messages matter: Effective messages for reaching tomorrow's innovators.* https://discovere.org/messages-matter/ (accessed March 20, 2024).

DiscoverE. 2024. *Engineers Week.* https://discovere.org/programs/engineers-week/ (accessed May 29, 2024).

DePass, A. L., and D. E. Chubin (eds.). 2015. *Understanding interventions that broaden participation in research careers, Vol. VI: Translating research, impacting practice.* Summary of a conference, San Diego, CA.

Innovative Research Group. 2017. *Public perceptions of engineers and engineering.* https://engineerscanada.ca/sites/default/files/public-perceptions-of-engineers-and-engineering.pdf (accessed April 15, 2024).

Jones, M. D., and D. A. Crow. 2017. How can we use the "science of stories" to produce persuasive scientific stories? *Palgrave Communications* 3(53) https://doi.org/10.1057/s41599-017-0047-7.

Lachney, M., and D. Nieusma. 2015. Engineering bait-and-switch: K-12 recruitment strategies meet university curricula and culture. *2015 ASEE Annual Conference & Exposition.* https://peer.asee.org/engineering-bait-and-switch-k-12-recruitment-strategies-meet-university-curricula-and-culture (accessed March 7, 2024).

Master, A., A. N. Meltzoff, and S. Cheryan. 2021. Gender stereotypes about interests start early and cause gender disparities in computer science and engineering. *Proceedings of the National Academy of Sciences* 118(48):e2100030118.

Margulies S. 2021. *Engineering impacts.* Presentation before the National Academies Committee on Extraordinary Engineering Impacts on Society. Washington, DC. September 27.

NAE (National Academy of Engineering). 2008. *Changing the conversation: Messages for improving public understanding of engineering.* Washington, DC: The National Academies Press. https://doi.org/10.17226/12187 (accessed April 15, 2024).

NAE. 2013. *Messaging for engineering: From research to action.* Washington, DC: The National Academies Press.

NAE/NRC (National Academy of Engineering and National Research Council). 2009. *Engineering in K–12 education: Understanding the status and improving the prospects.* Washington, DC: The National Academies Press. https://doi.org/10.17226/12635.

NAS/NAE/IOM (National Academy of Sciences, National Academy of Engineering, and Institute of Medicine). 2011. *Expanding underrepresented minority participation: America's science and technology talent at the crossroads.* Washington, DC: The National Academies Press. https://doi.org/10.17226/12984.

NASEM (National Academies of Sciences, Engineering, and Medicine) 2017. *Communicating science effectively: A research agenda.* Washington, DC: The National Academies Press. https://doi.org/10.17226/23674.

NASEM. 2019. *Minority serving institutions: America's underutilized resource for strengthening the STEM workforce.* Washington, DC: The National Academies Press. https://doi.org/10.17226/25257.

NCSES (National Center for Science and Engineering Statistics). 2023. *Diversity and STEM: Women, minorities, and persons with disabilities 2023.* Special Report NSF 23-315. Alexandria, VA: National Science Foundation. https://ncses.nsf.gov/wmpd (accessed April 15, 2024).

NSPE (National Society of Professional Engineers). 2024. *Engineers Week.* https://www.nspe.org/resources/partners-and-state-societies/engineers-week (accessed May 29, 2024).

Schank, G., and R. P. Abelson. 1995. Knowledge and memory: The real story. In Robert S. Wyer, Jr. (ed.), *Knowledge and memory: The real story.* Mahwah, NJ: Lawrence Erlbaum Associates. Pp. 1–85.

Scheufele, D. A. 2014. Science communication as political communication. *Proceedings of the National Academy of Sciences 111*(supplement_4):13585-13592.

Vest, C. 2011. The image problem for engineering: An overview. *The Bridge* 41(2):5–11.

Appendix A
Agenda – 2022 Symposium on Extraordinary Engineering Impacts on Society

Agenda – Thursday, August 18

10:00 a.m.   **Symposium greeting**
Dan Arvizu, Ph.D. – planning committee chair

10:05 a.m.   **Welcoming remarks**
John Anderson, Ph.D.
President, National Academy of Engineering

10:15 a.m.   **Overview of the symposium structure and logistics**
Dan Arvizu, Ph.D. – planning committee chair

SESSION I: NSF AND ITS ROLE IN FOSTERING
EXTRAORDINARY ENGINEERING IMPACTS ON SOCIETY

10:25 a.m.   **Introduction of session speakers**
Jeffrey Yost, Ph.D. – session moderator and planning committee member

10:30 a.m.   **The New Applied Innovation Policy Challenges Facing NSF**
William Boone Bonvillian, MAR, J.D.
Lecturer, Massachusetts Institute of Technology, Departments of Science,
Technology, and Society and Political Science; Senior Director, Special Projects,
MIT Office of Digital Learning.

10:50 a.m.   **The Influence of NSF Engineering Funding on Society**
Thomas Woodson, Ph.D.
Associate Professor and Graduate Program Director, Stony Brook University,
Department of Technology and Society

11:10 a.m.   **Moderated roundtable discussion – session speakers**

11:55 a.m.   **Break**

SESSION II: PEOPLE WHO BROUGHT ABOUT EXTRAORDINARY ENGINEERING IMPACTS ON SOCIETY

12:25 p.m.   **Introduction to the session**
             Ed Frank, Ph.D. – session moderator and planning committee member

12:30 p.m.   **Leadoff presentation**
             Pramod Khargonekar, Ph.D.
             Vice Chancellor for Research and Distinguished Professor of Electrical Engineering
                 and Computer Science, University of California, Irvine

1:00–3:20   **Session speakers**
p.m.   *Maia Weinstock, B.A. [speaking on Mildred Dresselhaus, Ph.D.]*
             author and Deputy Editorial Director, *MIT News,* Massachusetts Institute of
                 Technology

             Albert Pisano, Ph.D.
             Dean, Jacobs School of Engineering, University of California, San Diego

             Gilda Barabino, Ph.D.
             President and Professor of Biomedical and Chemical Engineering, Olin College of
                 Engineering

             Baratunde Cola, Ph.D.
             Professor, George W. Woodruff School of Mechanical Engineering, Georgia Tech

             Susan Estrada, B.S.
             Founder, CERFnet [retired]

             Paula Hammond, Ph.D.
             Institute Professor and Department Head of Chemical Engineering, Koch Institute of
                 Integrative Cancer Research, Massachusetts Institute of Technology

3:20 p.m.   **Break**

3:35 p.m.   **Moderated roundtable discussion – session speakers and discussant**
             <u>Discussant</u>
             Sarah Rajala, Ph.D.
             Professor Emeritus [E CPE]; James L. and Katherine S. Melsa Dean of Engineering,
                 Iowa State University

4:35 p.m.   **Session wrap-up and preview of Day 2**
             Dan Arvizu, Ph.D. – planning committee chair

4:45 p.m.   **Session adjourns**

10:00 a.m.   **Day 2 greeting; overview of the symposium structure and logistics**
Dan Arvizu, Ph.D. – planning committee chair

10:10 a.m.   **Keynote – Engineering's Role in Creating Extraordinary Impacts on Society**
Kristina Johnson, Ph.D.
President, The Ohio State University

10:30 a.m.   **Keynote – Engineering Education, The Key to Creating the Next Generation of Extraordinary Impacts**
Gary May, Ph.D.
Chancellor, University of California, Davis

SESSION III: NSF CENTERS THAT CATALYZED EXTRAORDINARY ENGINEERING IMPACTS ON SOCIETY

10:50 a.m.   **Introduction to the session**
Theresa Maldonado, Ph.D., PE – session moderator and planning committee member

11:00 a.m.   **Leadoff presentation**
Kon-Well Wang, Ph.D.
Stephen P. Timoshenko Collegiate Professor of Mechanical Engineering, University of Michigan

11:25 a.m.–   **Session speakers**
1:40 p.m.   *Mark Humayun, M.D., Ph.D.*
University Professor of Ophthalmology; Cornelius J. Pings Chair in Biomedical Sciences; Director, Institute for Biomedical Therapeutics; Co-Director USC Roski Eye Institute; and Director of Sensory Science Initiatives, Keck School of Medicine, University of Southern California

Karen Lozano, Ph.D., FNAI
Julia Beecherl Endowed Professor of Mechanical Engineering, and Founder and Director of the Nanotechnology Center, The University of Texas Rio Grande Valley

Gregory Deierlein, Ph.D.
John A. Blume Professor in the School of Engineering, Department of Civil and Environmental Engineering, Stanford University

Michael Silevitch, Ph.D.
Robert D. Black Professor and COE Distinguished Professor, Department of Electrical and Computer Engineering, Northeastern University

Veena Misra, Ph.D.
Distinguished Professor, Department of Electrical and Computer Engineering; and Director, Nanosystems Engineering Research Center on Advanced Self-Powered of Integrated Sensors and Technologies, North Carolina State University

S. Shankar Sastry, Ph.D. [accompanied by Larry Rohrbough, M.S.]
Director, Richard C. Blum Center for Developing Economies; Professor of Electrical
    Engineering and Computer Science, Professor of Bioengineering, and Thomas M.
    Siebel Professor, University of California, Berkeley

1:40 p.m.   **Moderated roundtable discussion – session speakers**

2:25 p.m.   **Break**

**SESSION IV: NSF PROCESSES THAT FOSTERED
EXTRAORDINARY ENGINEERING IMPACTS ON SOCIETY**

2:40 p.m.   **Introduction to the session**
             Louis Martin-Vega, Ph.D. – session moderator and planning committee member

2:50 p.m.   **Leadoff presentation**
             Andre Marshall, Ph.D.
             Vice President for Research, Innovation and Economic Development and President
                of the George Mason University Research Foundation

3:15–5:30   **Session speakers**
p.m.   *Rory A. Cooper, Ph.D.*
             Assistant Vice Chancellor, Distinguished Professor, and FISA Foundation and
                Paralyzed Veterans of America Professor of Rehabilitation Engineering,
                Department of Rehabilitation Science and Technology; Founding Director and
                VA Senior Research Career Scientist, Human Engineering Research Laboratories;
                University of Pittsburgh and U.S. Department of Veterans Affairs

             Harriet Nembhard, Ph.D.
             Dean, College of Engineering; Roy J. Carver Professor of Engineering; and Professor
                of Industrial Engineering; The University of Iowa

             Cindy Atman. Ph.D.
             Director and Lella Blanche Bowie Endowed Chair, Center for Engineering Learning
                & Teaching, and Professor, Human Centered Design & Engineering, University of
                Washington

             Sarah EchoHawk, M.N.M.
             Chief Executive Officer, American Indian Science and Engineering Society

             Alice Agogino, Ph.D.
             Chief Executive Officer and Co-Founder, Squishy Robotics; Roscoe and Elizabeth
                Hughes Professor of Mechanical Engineering, Emeritus and affiliated faculty with
                the Energy Resources Group, Women and Gender Studies, Development
                Engineering, and Studies in Science, Engineering and Mathematics Education;
                University of California, Berkeley

             Arlyne Simon, Ph.D.
             Solutions Architect, Health, Education, and Consumer Industries group, Intel;
                Founder, Abby Invents, LLC

5:30 p.m.   **Moderated roundtable discussion – session speakers and discussants**
<u>Discussants</u>
Mona Minkara, Ph.D.
Assistant Professor, Department of Bioengineering; Affiliated Faculty, Department of Chemistry and Chemical Biology; and Principal Investigator, COMBINE (Computational Modeling for BioInterface Engineering) Lab; Northeastern University College of Engineering

Sheri Sheppard, Ph.D., PE
Richard W. Weiland Professor in the School of Engineering, Emeritus, Stanford University

6:15 p.m.   **Symposium closing remarks and thanks**
Dan Arvizu, Ph.D. – planning committee chair

6:25 p.m.   **Symposium concludes**

REFERENCE

NASEM/NAE (National Academies of Sciences, Engineering, and Medicine and National Academy of Engineering). 2023. *Extraordinary engineering impacts on society: Proceedings of a symposium*. Washington, DC: The National Academies Press. https://doi.org/10.17226/26847.

Appendix B
Examples of Engineering Impacts on Society Outreach Materials[116]

EXAMPLE 1: MEET AN ENGINEER

Target Audience[s]

Young people (high school), particularly in historically marginalized groups in science, technology, engineering, and mathematics (STEM) such as underrepresented gender/racial/ethnic groups

Content Objectives

- Portray engineers as people who are social, creative, and have interests outside of work
- Demonstrate that engineers have shared identities and interests with the target audience
- Spur interest in engineering careers among students who are currently underrepresented in engineering through profiles of engineers who look like them and share their interests
- Connect engineering to awe and wonder, and spark curiosity
- Demonstrate that engineering is more than math and science skills and also involves problem solving and creativity

Media Format

Video-sharing social media platform (e.g., YouTube video) (2–5 minutes long)

Online Example Media Content*[117]

1.1 Meet an Engineer video.mp4 [link]
1.2 Transcript: Meet an Engineer video.pdf [link]
The example media content features Dr. Gary May, Chancellor, University of California, Davis.

Pitch Outline

Interview with a high-profile engineer who has received National Science Foundation (NSF) funding and who is from an underrepresented group, inspired by NSF's "Science Happens

---

[116] Content in this Appendix was prepared by staff of the Alan Alda Center for Communicating Science and the Massachusetts Institute of Technology under a contract with the National Academy of Sciences. The text has been reviewed and edited by the committee.

[117] Media examples are intended to depict the design language of a final product and are provided for reference only. They were created by the contractor and their content was not reviewed by the committee before production.

Here" campaign (NSF, 2023). Both the interviewee (the "subject") and interviewer should be from the target audience demographic. The interview will take place while doing something that is of interest to the engineer that may also appeal to the target audience (e.g., Dr. Gary May, chancellor of the University of California, Davis, and his interest in superheroes and his 13,000+ comic book collection; May, 2018).

Editing for this interview will contain elements about the specifics of this engineer's story but also connect to the bigger picture of what his or her story tells us about what it means to be an engineer and what it means to be a curious and creative human.

Script

- **Introduction of Video:** A voiceover (over a small montage of images) providing background information on the subject and interviewer. Then present additional information on the subject and the reasons why that person is being highlighted.
- **Insert a brief soundbite from the subject to stimulate interest in the conversation to follow.**
- **Interviewer:** "Today I had the honor of interviewing [subject], discussing [subject's] passion for their work and the future of engineering."
- **After the introduction, the edited interview will play.**

Examples of Interview Questions
- What was it that first piqued your interest in engineering?
- Based on most people's stereotypes about engineers, what do you think would surprise them the most about you?
- How are we working to make engineering more diverse, inclusive, equitable?
- What advice do you have for those that want to be engineers or work with them?
  - What about advice for people who don't usually see others who look like them in engineering? How can they get the support networks they need?
- Given unlimited money or time, what would you change about the world using engineering?

Rationale Behind Content Choices

In addition to being the right thing to do, there are also scientific and societal benefits to diversity—diversity leads to increased productivity (AlShelbi et al., 2018; Herring, 2009), better cognitive performance in collaborations (Freeman and Huang, 2015), and innovation gains (Hofstra et al., 2020). However, the engineering workforce is not representative of the richness and diversity of the U.S. population, which stifles innovation (Kozlowski et al., 2022; McGee, 2021). Women only make up 16 percent of college-educated engineered in the United States. In 2020, women, Black, and Latinx students received 24 percent, 5 percent, and 14 percent of engineering bachelor's degrees, respectively—well below these groups' proportional representation in the U.S. college-age population (NCSES, 2023).

Multiple complex factors lead to these disparities, but with this particular content, we seek to address those surrounding representation and exposure. Said simply, if you can't see it, you can't be it. This project features an interview with an engineer who shares interests and identities with the target audience in order to expand the images and representations of engineers that young people encounter. A previous study found that targeted messages and profiles of engineers who look like them, doing things they are interested in, can spur interest in engineering among underrepresented groups of young people (Discover Engineering, 2023).

A key science communication objective of this content is to demonstrate shared identities between the subject, interviewer, and the target audience. The tactics we use to reach this objective are designed according to best practices for strategic science communication (Besley and Dudo, 2022) and draw theoretically from social identity theory (Tajfel and Turner, 2004).

Another benefit of showcasing content in which the subject and interviewer come from the same underrepresented gender/racial/ethnic group as the target audience is that it also serves to challenge stereotypes about STEM careers and who can pursue them. Stereotype threats for Black students and for Asian students in STEM spaces provide simplistic, othering, views of students in STEM spaces that fail to see them in their full, complex humanity and psychologically burden them (McGee, 2018). Therefore, it is important that this content addresses stereotype threat.

An exemplar that we draw inspiration from for this content is The Story Collider, a nonprofit organization whose mission is "to reveal the vibrant role that science plays in all of our lives through the art of personal storytelling" (Story Collider, 2023). The Story Collider's commitment to representation in the science stories that are told on their stages works to challenge stereotypes (Neeley et al., 2020). Their theory of change is that through storytelling, people who do not fit the stereotype of what a scientists should be like can express themselves as full, complex humans to challenge this idea.

To build trust most effectively with our target audience, we draw from evidence-based science communication studies on trust. While most scientists and engineers think about trust as a unidimensional concept, it actually contains multiple dimensions: warmth, competence, being trusting/vulnerable, showing integrity, being willing to listen, and similarity to the target audience (Besley and Dudo, 2022; Colquitt and Salam, 2012; Fiske and Dupree, 2014). Because most engineers are perceived as high in competence but low in warmth, the dimension of trust we need to focus on most in this content is warmth (Fiske and Dupree, 2014). Fortunately, there are ways to build warmth perceptions that we can achieve through this "Meet an Engineer" video interview, including being personal, being curious, being transparent, being real, and demonstrating shared values (Besley and Dudo, 2022; Dixon et al., 2016; Fiske and Dupree, 2014). Overall, this interview's questions will be designed to demonstrate how engineers have good intentions and want to improve society in order to demonstrate the warmth dimension of trust (Besley and Dudo, 2022).

Finally, with this content we also want to consider culturally relevant ways to communicate. When it comes to interest, cultural background has a huge influence on the way that people approach STEM subjects (Davies and Horst, 2016). That is, it is not enough that both subject and interviewer are both members of the same cultural group as the audience. It is equally important to communicate in culturally relevant ways. Culture includes languages, customs, beliefs, knowledge, and the identity of a group or an individual, and all of these shape how they understand and make sense of the world. There is therefore a need to communicate in culturally relevant ways and to connect to the audience's culture (Medin and Bang, 2014) to make it more pertinent, understandable, and accessible—that is, to connect it to the reality of their everyday lives.

Example 1 References

AlShebli, B. K., T. Rahwan, and W. L. Woon. 2018. The preeminence of ethnic diversity in scientific collaboration. *Nature Communications* 9(1):5163.

Besley, J. C., and A. Dudo. 2022. *Strategic science communication: A guide to setting the right objectives for more effective public engagement.* Baltimore, MD: Johns Hopkins Press.

Colquitt, J. A., and S. C. Salam. 2012. Foster trust through ability, benevolence, and integrity. In E. A. Locke (ed.), *Handbook of principles of organizational behavior: Indispensable knowledge for evidence-based management.* Hoboken, NJ: John Wiley & Son. Pp. 389–404.

Davies, S. R., and M. Horst. 2016. *Science communication: Culture, identity and citizenship.* New York: Springer.

Discover Engineering. 2023. *Messages matter: Effective messages for reaching tomorrow's innovators.* [White paper]. https://discovere.org/wp-content/uploads/2023/02/DiscoverE-Messages-Matter-Executive-Summary-Feb-2023.pdf (accessed April 23, 2024).

Dixon, G., K. McComas, J. Besley, and J. Steinhardt. 2016. Transparency in the food aisle: The influence of procedural justice on views about labeling GM foods. *Journal of Risk Research* 19(9):1158–1171.

Fiske, S. T., and C. Dupree. 2014. Gaining trust as well as respect in communicating to motivated audiences about science topics. *Proceedings of the National Academy of Sciences* 111(Suppl 4):13593–13597.

Freeman, R. B., and W. Huang. 2015. Collaborating with people like me: Ethnic coauthorship within the United States. *Journal of Labor Economics* 33(S1):S289–S318.

Herring, C. 2009. Does diversity pay?: Race, gender, and the business case for diversity. *American Sociological Review* 74(2):208-224.

Hofstra, B., V. V. Kulkarni, S. Munoz-Najar Galvez, B. He, D. Jurafsky, and D. A. McFarland. 2020. The diversity–innovation paradox in science. *Proceedings of the National Academy of Sciences* 117(17):9284–9291.

Kozlowski, D., V. Larivière, C. R. Sugimoto, and T. Monroe-White. 2022. Intersectional inequalities in science. *Proceedings of the National Academy of Sciences* 119(2):e2113067119.

May, G. S. 2018. *Chancell-ing: Treating my action heroes to "Black Panther."* https://www.ucdavis.edu/news-chancelling-treating-action-heroes-black-panther (accessed February 2, 2024).

McGee, E. 2018. "Black genius, Asian fail": The detriment of stereotype lift and stereotype threat in high-achieving Asian and Black STEM students. *AERA Open* 4(4):2332858418816658.

McGee, E. O. 2021. *Black, brown, bruised: How racialized STEM education stifles innovation.* Cambridge, MA: Harvard Education Press.

Medin, D. L., and M. Bang. 2014. The cultural side of science communication. *Proceedings of the National Academy of Sciences* 111(Suppl 4):13621–13626.

NCSES (National Center for Science and Engineering Statistics). 2023. *Diversity and STEM: Women, minorities, and persons with disabilities 2023.* Special Report NSF 23-315. Alexandria, VA: National Science Foundation. https://ncses.nsf.gov/pubs/nsf23315/ (accessed March 9, 2023).

NSF (National Science Foundation). 2024. *Science Happens Here.* https://new.nsf.gov/about/science-happens-here (accessed February 2, 2024).

Neeley, L., E. Barker, S. R. Bayer, R. Maktoufi, K. J. Wu, and M. Zaringhalam. 2020. Linking scholarship and practice: Narrative and identity in science. *Frontiers in Communication* 5:35.

Saffran, L., S. Hu, A. Hinnant, L. D. Scherer, and S. C. Nagel. 2020. Constructing and influencing perceived authenticity in science communication: Experimenting with narrative. *PLOS One* 15(1):e0226711.

Story Collider. 2024. *About the Story Collider*. https://www.storycollider.org/about-us-2 (accessed February 2, 2024).

Tajfel, H., and J. C. Turner. 2004. The social identity theory of intergroup behavior. In J. T. Jost and J. Sidanius (eds.), *Political Psychology*. Oxfordshire, England: Psychology Press. Pp. 276–293.

EXAMPLE 2: QUEEN OF CARBON FAMILY TREE

Target Audience[s]

High school students (especially women)

Content Objectives

- Frame basic engineering research as something that has far-reaching effects in the real world
- Counter the misperception of the engineer as a lone genius by telling the story of someone who was influenced by—and in turn influenced—many, thus framing engineering as a social endeavor
- Promote gender equity in the histories we tell of engineering

Media Format

Clickable, interactive graphic (web) post

Online Example Media Content[*]

2.1 - Carbon Queen – Windows executable file.exe [link]
2.2 - Carbon Queen Prezi presentation format embed code.pdf [link]
2.3 - Carbon Queen script.pdf [link]
2.4 - Carbon Queen - Apple OS-formatted files.zip [link]

Pitch Outline

The content is illustrated in a clickable graphic containing the many forms of carbon connected in a "family tree"-type diagram illustrating their real-world applications. Images of the engineers and scientists who work with these materials are entangled in the tree graphic to illustrate that engineering is a network and not the domain of a single genius. This tree celebrates the many contributions that Dr. Millie Dresselhaus has made to engineering (Weinstock, 2023), highlighting those supported by NSF. It also illustrates the amazing "academic family tree" that has resulted from her work.

The interactive graphic example contains some, but not all, of the content proposed in the script outline.

Script Outline

We see a woman—a representation of Millie Dresselhaus—sitting under a tree writing with a pencil. When you first click on it, it zooms in on a pencil and presents a short blurb about graphite, introducing the reader to carbon and its many uses. The story of how Dresselhaus was initially inspired by carbon is then introduced in the form of a blurb and picture of her that pops up.

We then move to roots of the tree. Every part of a tree stores carbon, from the roots to the trunk, branches, and leaves. Carbon stored underground, in particular, is considered to be the most stable reservoir for the element. Graphite intercalation compounds are addressed here; leading to insights on how the properties of graphite can be altered.

Moving up the tree, branches highlight other researchers and the technologies that have "branched off" from the roots, such as buckyballs and flexible computer screens and quantum computers made from graphene.

The tree's vascular system represents carbon fibers and carbon nanotubes, leading to such technologies as organic light-emitting diode (OLED) displays and nanocarriers used to transport chemotherapy agents. Tree leaves represent fuel cells; droplets on those leaves introduce the concept of water purification through nanofiltration.

Integrated into these vignettes is mention of how Dresselhaus's research laid the groundwork for all of these innovations. This includes specific callouts to the two researchers—Richard Smalley and Andre Geim—who acknowledged her in their Nobel Prize lectures.

Going back to the woman doodling under a tree, the user might click on the notebook she's writing in. This contains a small cartoon of her with her four children and small blurb on her accomplishments specifically as a woman in engineering just to close out the presentation, but this should not be the focus of the piece as the intent is to honor her foundational contributions, not that she achieved them "despite" being a woman.

Rationale Behind Content Choices

It matters how we tell the history of engineering. While stories of science and engineering often portray it as a white, Western, and male-dominated domain (Rasekoala and Orthia, 2020), in truth, more diverse voices have always been involved (Finlay et al., 2021).

Studies indicate that girls' enthusiasm for STEM subjects tends to decline during middle school (U.S. Department of Education, 2006). Research suggests that girls are often less interested in subjects like computer science or engineering that are characterized by gender stereotypes, suggesting a need to better disrupt these stereotypes early in education (Master et al., 2021). Engineering is stereotyped in modern American culture as a male-oriented field that involves social isolation, an intense focus on machinery, and innate brilliance (Cheryan et al., 2015).

An opportunity exists to expand the image of what it means to be an engineer beyond these narrow stereotypes. Broadening the representation of the people who do this work, the work itself, and the environments in which it occurs has been shown to significantly increase girls' sense of belonging and interest in the field (Cheryan et al., 2015). As Cheryan and colleagues remark, "Rather than attempting to overhaul current stereotypes, which may deter some men and women, a more effective strategy may be to diversify the image of these fields so that students interested in these fields do not think that they must fit a specific mold to be a successful . . . engineer" (p. 6).

One effective way to target girls in messaging is to elevate the biographies of female engineers (Discover Engineering, 2023). This interactive image-forward blog post about Millie Dresselhaus's "family tree" of collaborators and innovations deriving from her discoveries aims to not only elevate her biography but also to subvert some common narratives about engineering. Typically, the stories of engineering have been told as glossy hero narratives that frame stories of achievement as individual endeavors (Davies, 2021; Fahy, 2015; Felt and Fochler, 2013). But engineering is a communal, not individual; engineering is a social process (Kuhn, 1962; Oreskes, 2021). This blog post, in showing those who supported Millie and those she supported, demonstrates the social connectivity of not only her story but of the work that NSF funds.

The use of visuals in this blog post is intended to engage audiences who may not already be interested or motivated to learn about engineering. For example, a study using the Elaboration

Likelihood Model (ELM) found that audiences engaged more deeply with climate change issues that were presented with infographics than those presented with either text-only or illustration, with their learning preferences and visual literacies moderating the effect (Lazard and Atkinson, 2015). The ELM describes how attitudes change as a result of persuasive messaging either through central (deeper engagement with content) or peripheral processing (mental shortcuts), depending on whether an individual is highly informed and motivated about the topic (central) or not (peripheral) (Petty and Cacioppo, 1986). The audience for this content is likely more on the "peripheral" processing mode, so visuals may be the most effective way to engage them.

However, studies also suggest that a poor-quality picture is worse than no picture at all (Zhu et al., 2021). It is thus important to apply the best practices from the field of visual design, especially (1) treating visuals as integrated and not an add-on and (2) clearly identifying target audiences and refining visuals for them specifically (Rodríguez Estrada and Davis, 2015). Visuals must also be accessible, with such tools such as alt text, clear captions, and color contrast incorporated into the media (Crameri et al., 2020; Schwabish et al., 2022).

Example 2 References

Cheryan, S., A. Master, and A. N. Meltzoff. 2015. Cultural stereotypes as gatekeepers: Increasing girls' interest in computer science and engineering by diversifying stereotypes. *Frontiers in Psychology* 6:49.

Crameri, F., G. E. Shephard, and P. J. Heron. 2020. The misuse of colour in science communication. *Nature Communications* 11(1):5444.

Davies, S. R. 2021. Performing science in public: Science communication and scientific identity. *Community and Identity in Contemporary Technosciences* 31:207.

Discover Engineering. 2023. *Messages matter: Effective messages for reaching tomorrow's innovators.* [White paper]. https://discovere.org/wp-content/uploads/2023/02/DiscoverE-Messages-Matter-Executive-Summary-Feb-2023.pdf (accessed April 24, 2024).

Fahy, D. 2015. *The new celebrity scientists: Out of the lab and into the limelight.* Lanham, MD: Rowman and Littlefield.

Felt, U., and M. Fochler. 2013. What science stories do: Rethinking the multiple consequences of intensified science communication. In P. Baranger and B. Schiele (eds.), *Science communication today. International perspectives, issues and strategies.* Paris: CNRS Editions. Pp. 75-90.

Finlay, S. M., S. Raman, E. Rasekoala, V. Mignan, E. Dawson, L. Neeley, and L. A. Orthia. 2021. From the margins to the mainstream: Deconstructing science communication as a white, Western paradigm. *Journal of Science Communication* 20(1):C02.

Kuhn, T. S. (1962/1996. *The structure of scientific revolutions* (3rd ed). Chicago: The University of Chicago Press.

Lazard, A., and L. Atkinson. 2015. Putting environmental infographics center stage: The role of visuals at the elaboration likelihood model's critical point of persuasion. *Science Communication* 37(1):6.

Master, A., A. N. Meltzoff, and S. Cheryan. 2021. Gender stereotypes about interests start early and cause gender disparities in computer science and engineering. *Proceedings of the National Academy of Sciences* 118(48):e2100030118.

Oreskes, N. 2021. *Why trust science?* Princeton, NJ: Princeton University Press.

Petty, R. E., and J. T. Cacioppo. 1986. The elaboration likelihood model of persuasion. In L. Berkowitz (ed.), *Advances in experimental social psychology, vol. 19.* Cambridge, MA: Academic Press. Pp. 123–205.

Rasekoala, E., and L. Orthia. 2020. Anti-racist science communication starts with recognising its globally diverse historical footprint. *Impact of Social Sciences Blog*, July 1. https://blogs.lse.ac.uk/impactofsocialsciences/2020/07/01/anti-racist-science-communication-starts-with-recognising-its-globally-diverse-historical-footprint/ (accessed April 24, 2024).

Rodríguez Estrada, F. C., and L. S. Davis. 2015. Improving visual communication of science through the incorporation of graphic design theories and practices into science communication. *Science Communication* 37(1):140–148.

Schwabish, J., S. J. Popkin, and A. Feng. 2022. *Do no harm guide: Centering accessibility in data visualization.* Urban Institute. Available at https://www.urban.org/research/publication/do-no-harm-guide-centering-accessibility-data-visualization (accessed April 23, 2024).

U.S. Department of Education. 2006. *The condition of education.* Washington, DC: National Center for Education Statistics, U.S. Government Printing Office.

Zhu, L., L. S. Davis, and A. Carr. 2021. A picture is not always worth a thousand words: The visual quality of photographs affects the effectiveness of interpretive signage for science communication. *Public Understanding of Science* 30(3):258–273.

EXAMPLE 3: EARTHQUAKE SHAKE TABLE

Target Audience[s]

General public (teens and up), especially those who live in high hazard earthquake areas of the United States

Content Objectives

- Make visible some of the innovations behind the innovation, and how behind-the-scenes investments from the NSF lead to big societal impacts
- Inspire awe about the "cool-factor" of the United States' largest shake table
- Frame engineering as important to disaster and hazard mitigation

Media Format

TikTok "Did You Know?" format or short (2 minute) YouTube video

Online Example Media Content*

3.1 - Earthquake Shake Table - video.mp4 [link]

Pitch Outline

Start by pointing out an innocuous structure on a building and pointing out that it actually makes the building safer from earthquakes. (Cut in a video of a push-puppet as a metaphor for how the base-isolation mechanism works.)

"There is no doubt that as our planet continues to evolve, extreme weather events, like severe storms, floods, earthquakes, and wildfires are likely to cause unprecedented damage to infrastructure, communities, and ecosystems," said Sethuraman Panchanathan (NSF director). "As we look to build a more resilient country, we must continue to leverage our science and engineering expertise to expand prosperity for our nation's communities and protect the critical infrastructure that supports them."

Currently, NSF is the only federal agency that supports research across all fields of science, technology, engineering, and medicine and all levels of STEM education, a vitally important distinction when it comes to building resilient futures and mitigating natural hazards.

Earthquakes are a danger to buildings and lives. Engineers are deeply concerned about designing and retrofitting buildings so they can be safer. There are many factors at play, from design aesthetics to zoning regulations and to keeping buildings affordable. For example, as we try to mitigate climate change, we also want to use less concrete, which is carbon intensive, and that leads us to explore other materials like wood (which is a carbon sink).

So, what do we do? Well, here is the story behind how the largest (in the United States) height capacity three-dimensional shake-table was developed. Engineers can simulate buildings on a computer, but we need a good way to experimentally test this in a way that won't destroy real buildings or hurt people. With this table, engineers can mimic past earthquake conditions and reproduce their movement. This can be used to make older building stronger and to help with new design standards for new buildings.

Back to the image of the innocuous-looking piece of a building. So, next time you see this, you can feel a little bit safer, thanks to a very "shaky" table.

Optionally, a follow-up video would feature a presenter who creates his or her own structure and then tests it on a shake table.

Rationale Behind Content Choices

This content is meant for a general audience not already motivated and interested in engineering. According to the Elaboration Likelihood Model,[118] attitudes change as a result of persuasive messaging through one of two pathways: central (deeper) or peripheral (shallower). Central processing involves more effortful thinking, and attitudes formed via this route are longer-lasting and more resistant to change (Petty and Cacioppo, 1986). Peripheral processing, on the other hand, is less effortful, surface-level, subject to mental shortcuts, and does not result in attitudes that are as enduring or resilient (Petty and Cacioppo, 1986). This implies that with an audience that is not already necessarily aware or interested in engineering, we will need to rely more on heuristic cues than information to engage them (Scheufele and Turney, 2006).

A study of the most popular science TikTok videos found that this genre is dominated by entertaining physics or chemistry experiments, indicating that users engage with getting to see the process of science at work on this platform ("procedural science") (Zeng et al., 2020). This is especially important on a platform like TikTok, which has an algorithm that allows for the discovery of content beyond the user's usual social network. TikTok's user base is majority young and female-identifying, a population that is currently underrepresented in engineering. Using a short-duration video to explain procedurally a "cool" piece of engineering technology or infrastructure may thus be an effective way to stimulate greater interest in the field in this group.

While videos are increasingly used in science communication, they can be less effective than a slideshow if they are too long (Iamamura et al., 2020). In terms of remembering scientific knowledge, the best types of videos seem to be either narrative explanatory or animated; both had scientific content remembered much better than those in expert films (like those favored by research institutions and universities) (Boy et al., 2020). While the objectives with this content are not necessarily to teach, the narrative explanatory style is still likely to be more effective than a formal video. For this reason, a more informal, "popular" YouTube or TikTok style video would seem appropriate for reaching these specific audiences.

The example content is intended to spur interest, curiosity, wonder, and awe in the audience through the earthquake shake table.[119] There are a variety of types of awe used in science communication (Luna and Bering, 2020), but awe primarily works when an audience encounters something that affects them emotionally and shifts their worldview or motivates curiosity. At the same time, there is a need to be careful about burning people out on constantly "insisting on sparking interest and curiosity" because this is an exhausting emotion to maintain for long periods of time (Davies, 2019). That is, there are multiple reasons to keep this video short in length.

With such a short video, it will also be important to be intentional about narrative design, choosing a single topic to address (rather than a bullet point list) and structuring the presentation as a story rather than a list of facts (Baron, 2010; Olson, 2015). Additionally, research suggests

---

[118] The Elaboration Likelihood Model is also addressed in the rational section of the Content 2 discussion.
[119] Note that the *PBS Kids* website has a create-your-own-shake-table and a test-a-structure instructional ("Design Squad Global – Seismic Shake-up") aimed at elementary school -aged children: https://pbskids.org/designsquad/build/seismic-shake-up/.

that an "infotainment" narration style—which uses humor, personality, and informal language—is more appropriate for a less formally educated audience and better overall for learning than an expository narration style that uses formal language and conveys authority (Davis et al., 2020).

Example 3 References

Baron, N. 2010. *Escape from the ivory tower: A guide to making your science matter.* Washington, DC: Island Press.

Boy, B., H. J. Bucher, and K. Christ. 2020. Audiovisual science communication on TV and YouTube. How recipients understand and evaluate science videos. *Frontiers in Communication* 5:608620.

Davies, S. R. 2019. Science communication as emotion work: Negotiating curiosity and wonder at a science festival. *Science as Culture* 28(4):538–561.

Davis, L. S., B. León, M. J. Bourk, and W. Finkler. 2020. Transformation of the media landscape: Infotainment versus expository narrations for communicating science in online videos. *Public Understanding of Science* 29(7):688-701.

Imamura, K., K. T. Takano, N. H. Kumagai, Y. Yoshida, H. Yamano, M. Fujii, T. Nakashizuka, and S. Managi. 2020. Valuation of coral reefs in Japan: Willingness to pay for conservation and the effect of information. *Ecosystem Services* 46:101166.

Luna, D. S., and J. M. Bering. 2021. The construction of awe in science communication. *Public Understanding of Science* 30(1):2–15.

Olson, R. 2015. Houston, we have a narrative. In *Houston, we have a narrative.* Chicago: University of Chicago Press.

Petty, R. E., and J. T. Cacioppo. 1986. The elaboration likelihood model of persuasion. In *Communication and Persuasion* (pp. 1-24). Springer: New York, NY.

Scheufele, D. A. 2006. Messages and heuristics: How audiences form attitudes about emerging technologies. In J. Turney (ed.), *Engaging science: Thoughts, deeds, analysis, and action.* London, UK: The Wellcome Trust. Pp. 20–25. https://cspo.org/legacy/library/090423F3NZ_lib_ScheufeleDA2006M.pdf (accessed February 5, 2024).

Weinstock, M. 2023. *Carbon queen.* Cambridge, MA: The MIT Press.

Zeng, J., M. S. Schäfer, and J. Allgaier. 2020. Reposting "till Albert Einstein is TikTok famous": The memetic construction of science on TikTok. *International Journal of Communication* 15:3216–3247.

EXAMPLE 4: GRAND CHALLENGES IN ENGINEERING

Target Audience[s]

Elementary and middle school students

Example Objectives

- Debunk the lack of representation in engineering as only a "pipeline" problem
- Encourage culturally relevant pedagogy in STEM content
- Encourage young people to bring their whole selves (especially culturally) into STEM spaces

Media Format

Workbook style, fill-in-the-blank booklet aimed at elementary school students; blog posts aimed at middle school students

Online Example Media Content[*]

4.1 - Grand Challenges in Engineering illustrative blog post.pdf [link]

Pitch Outline

The workbook will show the reader that he or she can already think like an engineer. Using a fill-in-the-blank format, it will have the reader engage in some design steps using drawings.

1) Name a problem that you already care about. Put it in this title box.
2) Next, draw a device that could help you solve that problem (box provided for drawing).
3) What are some good things about this device? Does it save money? Power? Help the environment? Show this on your drawing.
4) What skills might you need to make this device? (art, math, science, ….)
5) What other types of people might you need to do this? (teachers, construction workers, doctors, artists, …)

See! You are ALREADY an engineer, designing solutions to problems in the world around you.

Perhaps a means might be implemented—via a QR code, for example—that would allow the students' drawings to be posted anonymously (after screening) to an NSF public website where staff could connect the contribution to NSF-funded efforts.

The workbook would finish with encouragement to the students to continue applying engineering thinking in their lives and links where they could learn more.

The blog posts will introduce some of the grand challenges that the world is facing (e.g., climate change, health, access to clean water and food, keeping our computers safe). Using visuals of people and featuring those who are underrepresented in engineering and who are working on some of these challenges (to increase feelings of self-efficacy), it will introduce the notion of engineers tackling the world's biggest problems. This will be accomplished by first showing past innovations brought about by engineers, such as automobiles, aircraft, telephones, and computers. It will then segue into more recent inventions such as artificial retinas and robotics to enhance the operation and versatility of wheelchairs—two technologies brought about in part by NSF funding—and the people behind them (NASEM, 2023). Examples will be chosen to demonstrate that engineering is richer when it is more diverse and that engineers use their own backgrounds and points of view to make the world a better place.

Rationale Behind Content Choices

Communication scholarship informs us that simply sharing information does not change hearts and minds (i.e., beliefs and attitudes) (Allum et al., 2008; Sturgis and Allum, 2004). However, much of our messaging about engineering careers uses this approach.

Instead, this content uses "problem posing" methodology, drawing from the *Pedagogy of the Oppressed* monograph (Freire, 2020), asking students a question and having them work with that rather than giving them answers (Morgan and Saxton, 2006). This allows the young person to take on the "mantle of the expert," a technique that comes from drama-based pedagogy (Dawson and Lee, 2018; Wilhelm, 2002). The benefits of using the mantle-of-the-expert approach are that young people feel respected by having expert status and insights and gain an understanding of the occupation they are exploring. This strategy draws on participants' interests and may motivate them to research more about engineering topics and careers.

Additionally, this content draws upon young people's funds of knowledge and helps them draw culturally relevant connections to the field of engineering. "Funds of knowledge" describes the accumulated life experiences, skills, and knowledge used to navigate everyday social contexts in the homes of students of color (González and Moll, 2002).

Art-based engineering communication like this content engages multiple senses and supports the young person in thinking creatively, which has been shown to lead to better learning outcomes in science topics (Jacobson et al., 2016). That is, while visuals for science communication can be effective, visualizing can also be a very effective learning tool for STEM topics (Evagorou et al., 2015). This also embraces multimodal ways of learning, which support more accessible communications through the Universal Design for Learning teaching approach (CAST, 2024).

Positioning audience members as holders of knowledge asks them to use their imagination, which puts young people in charge of their own learning and meaning-making (Greene, 1995). This may also be helpful in developing their engineering or science identity, which is thinking of oneself and being recognized by others as being an engineering or science person (Carlone and Johnson, 2007; Godwin, 2016). Research on science/engineering identity reveals there are three dimensions to the concept: competence, performance, and recognition, which this content aims to support (Kim and Sinatra, 2018).

This content also seeks to increase the audience's feelings of self-efficacy in engineering by engaging in ideation around design (e.g., that they can do it and have the skills and imagination to do so) (Bandura, 2002). It also frames engineering as something that is creative and not just about skills in math and science. And the content helps support a sense of response efficacy, that engineering is something that has the potential to make a difference in the world. Research shows that underrepresented students are more likely to pursue STEM research for community-minded reasons (Spalter-Roth and Van Vooren, 2009) and recruiting these groups into these spaces requires emphasizing the ways in which engineering and other STEM subjects contribute to local communities.

The concept of efficacy is related to a number of communication models that seek to change behavior (e.g., pursuing an engineering career). For example, the integrated behavior model predicts that people's attitude, norms, and efficacy beliefs shape their intentions and eventually their behaviors (Ajzen, 1991). In this case, the hoped-for outcome is to stimulate young people to study engineering by first becoming interested in it enough to look more into it. Self efficacy can be built both by example (e.g., through messages) and through direct experience (Bandura, 1977). This content aims to do both, with a focus on direct experience.

Solving society's wicked problems will require embracing multiple forms of knowing, not only from engineering but also from beyond engineering (Medin and Bang, 2014). We live in a complex world where we must make important decisions with incomplete evidence and account for others' values (Funtowicz and Ravetz, 2018). This means our messaging should demonstrate the ways in which engineering is already aligned with young people's interests and wish to help their communities. The content seeks to go beyond just broadening participation in engineering to also welcoming and empowering young people in these spaces (Bevan et al., 2020; Dawson, 2019; Humm and Schrögel, 2020; YESTEM Project UK Team, 2020).

Example 4 References

Ajzen, I. 1991. The theory of planned behavior. *Organizational Behavior and Human Decision Processes* 50(2):179–211.

Allum, N., P. Sturgis, D. Tabourazi, and L. Brunton-Smith. 2008. Science knowledge and attitudes across cultures: A meta-analysis. *Public Understanding of Science* 17(1):35–54.

Bandura, A. 1977. Self-efficacy: Toward a unifying theory of behavioral change. *Psychological Review* 84(2):191–215. https://doi.org/10.1037/0033 -295X.84.2.191.

Besley, J.C., and A. Dudo. 2022. *Strategic science communication: A guide to setting the right objectives for more effective public engagement*. Baltimore, MD: Johns Hopkins Press.

Bevan, B., A. C. Barton, and C. Garibay. 2020. Broadening perspectives on broadening participation: Professional learning tools for more expansive and equitable science communication. *Frontiers in Communication* 5:52.

Carlone, H. B., and A. Johnson. 2007. Understanding the science experiences of successful women of color: Science identity as an analytic lens. *Journal of Research in Science Teaching* 44(8):1187–1218.

CAST. 2024. *About universal design for learning*. http://www.cast.org/our-work/about-udl.html (accessed April 23, 2024).

Dawson, E. 2019. *Equity, exclusion and everyday science learning: The experiences of minoritised groups*. New York: Routledge.

Dawson, K., and B. K. Lee. 2018. *Drama-based pedagogy: Activating learning across the curriculum*. Bristol, UK: Intellect Books.

Evagorou, M., S. Erduran, and T. Mäntylä. 2015. The role of visual representations in scientific practices: From conceptual understanding and knowledge generation to "seeing" how science works. *International Journal of STEM Education* 2(1):1–13.

Funtowicz, S., and J. Ravetz. 2008. Post-normal science. In N. Castree, M. Hulme, and J. D. Proctor (eds.), *Companion to environmental studies*. New York: Routledge. Pp. 443–447.

Godwin, A. 2016, January. The development of a measure of engineering identity. In *ASEE Annual Conference and Exposition*. https://doi.org/10.18260/p.26122. https://par.nsf.gov/biblio/10042227 (accessed April 23, 2024).

González, N., and L.C. Moll. 2002. Cruzando el puente: Building bridges to funds of knowledge. *Educational Policy (Los Altos, Calif.)* 16(4):623–641.

Greene, M. 1995. *Releasing the imagination: Essays on education, the arts and social change*. San Francisco: Josey-Bass. Inc.

Freire, P. 2020. Pedagogy of the oppressed. In J. Beck, C. Jenks, N. Keddie, and M. F. D. Young (eds.), *Toward a sociology of education*. New York: Routledge. Pp. 374-386

Humm, C., and P. Schrögel. 2020. Science for all? Practical recommendations on reaching underserved audiences. *Frontiers in Communication* 5:42.

Jacobson, S. K., J. R. Seavey, and R. C. Mueller. 2016. Integrated science and art education for creative climate change communication. *Ecology and Society* 21(3):26269971.

Kim, A.Y., and G. M. Sinatra. 2018. Science identity development: An interactionist approach. *International Journal of STEM Education* 5(1):1–6.

Medin, D. L., and M. Bang. 2014. The cultural side of science communication. *Proceedings of the National Academy of Sciences* 111 (Suppl 4):13621–13626.

Morgan, N., and J. Saxton. 2006. *Asking better questions*. Markham, ON: Pembroke Publishers Limited.

NASEM (National Academies of Sciences, Engineering, and Medicine). 2023. *Extraordinary engineering impacts on society: Proceedings of a symposium*. Washington, DC: The National Academies Press. https://doi.org/10.17226/26847.

Spalter-Roth, R., and N. Van Vooren. 2009. Idealists vs. careerists: Graduate school choices of sociology majors. American Sociological Association Department of Research and Development. https://www.asanet.org/wp-content/uploads/files/pdf/ideaslistcareerist.pdf (accessed April 23, 2024).

Sturgis, P., and N. Allum. 2004. Science in society: Re-evaluating the deficit model of public attitudes. *Public Understanding of Science* 13(1):55–74.

Wilhelm, J. D. 2002. *Action strategies for deepening comprehension*. New York: Scholastic Inc.

YESTEM Project UK Team. 2020. The equity compass: A tool for supporting socially just practice. https://yestem.org/wp-content/uploads/2020/10/EQUITY-COMPASS-YESTEM-INSIGHT.pdf (accessed April 23, 2024).

EXAMPLE 5: EXTRAORDINARY IMPACTS OF ENGINEERING

Target Audience

Gen Z/younger millennials

Content Objective

Framing engineering innovations as beneficial to people's everyday lives, rather than just being "out there"

Media Format

TikTok/Reel-style video

Online Example Media Content*

5.1 - Extraordinary Impacts of Engineering video.pdf [link]
5.2 – Transcript—Extraordinary Impacts of Engineering video.pdf [link]

Pitch Outline

This content will feature a TikTok/Reel-style video that incorporates nostalgia and concise, descriptive language to engage the audience in a brief video format. Our aim is to create a video that will not only capture the attention of viewers but also encourage them to watch it multiple times. To achieve this, we plan to make the video vivid, humorous, and, of course, interesting.

I. Introduction
- The video aims to offer viewers a unique perspective by showcasing life without some of our favorite pieces of technology, such as smartphones.

II. Keeping the Video Short
- Shorter videos encourage more views, which increases engagement, and improves the video's visibility on TikTok's algorithm. Essentially, keeping the video short maximizes our chances of reaching a wider audience and boosting the video's success on the platform.

III. Incorporating Nostalgia
- To successfully target Gen Z and millennials, the video use nostalgia and concise language to immediately grab their attention. Connecting with them emotionally, especially through the highly regarded 2000s nostalgia will contribute significantly to the video's success.

Video Concept

1. A journey through time: Short clips from different decades transition seamlessly, showcasing people using payphones, film and digital cameras, address books, physical planners, and calendars in settings that have the appropriate aesthetic. The aim is to highlight how far we've come with technology and how it has revolutionized our lives. The video ends with a clip of the smartphone, a testament to

the ingenuity of engineers and the incredible advancements in technology that have taken place.

2. Animation: This could include an educational video in the form of a funny animation. The purpose of including an animation TikTok is that it ties beautifully with nostalgia as Gen Z and millennials grew up with educational animations such as the Disney Channel Healthy Handbook video,[120] BrainPOP,[121] and Wild Kratts.[122]

Script

As noted, the video will be a compilation of different videos sourced from other videos where we see people using payphones, film, and digital cameras, etc. followed by a voiceover that will begin when the smartphone is introduced: "Over the years, we have witnessed incredible technological advancements, all thanks to the unwavering efforts of engineers."

Figure B-1 illustrates a rough draft quick concept for how the video will transition.

FIGURE B-1 Rough draft illustrating the potential transition of vignettes in the text example "Extraordinary Engineering Impacts" video.

The voiceover continues, in the style of a "Did you know?" TikTok video:

"Hey, did you know that almost everything we use on a daily basis is a gift from our engineering heroes?

"Take the smartphone, for example. It has revolutionized our lives in ways we never thought possible. Remember the days when we had to carry a bulky camera, planner, address book, contacts, and even a coin purse just to use a payphone? Phew, talk about a workout! But now, we can fit all of that and more in our jeans pocket, thanks to the mighty smartphone."

"So, let's give a big round of applause to engineers for making our lives easier, and our pockets lighter!"

The humor takes its inspiration from 1980s commercials, which should strike a chord with the Gen Z and millennial audience.

If a more direct tie-in to NSF-funded engineering innovations is desired, then any of the more recent examples of these noted in the symposium proceedings or report can be used as the topic and mentioned in the script.

---

[120] https://www.tiktok.com/@disneycommercials/video/7216056657204219179.
[121] https://www.youtube.com/channel/UCJ5dVwsCLKlWuOZyi7WDwfw.
[122] https://pbskids.org/wildkratts/videos/.

Rationale Behind Content Choices

Almost every object we use in our daily lives comes to us through the work of engineers, and numerous innovations can be tied back to National Science Foundation funding. However, research shows that most Americans do not know what engineers do or make the connection between engineer's work and these innovations (Innovative Research Group, 2017; NAE, 2008). While they see engineers as a profession with a high level of expertise (competence), they do not connect the profession to its immense social value. Therefore, this content seeks to connect engineering innovations that affect people's daily lives in a humorous way, by challenging people to imagine their lives without a piece of technology.

This content uses a persuasion technique known as "emphasis framing" which encourages an audience to think about an issue in a specific way (Druckman, 2001). The technique "select[s] some aspects of a perceived reality and make[s] them more salient in a communicating text, in such a way as to promote a particular problem definition, causal interpretation, moral evaluation, and/or treatment recommendation" (Entman, 1993, p. 52).

The messaging in this content draws from branding studies in engineering that found adults and teens of all genders rated "Engineers make a world of difference" as the most appealing marketing message for general audiences (NAE, 2008). It also draws from work in communication scholarship that suggests framing learning about science as an everyday experience rather than something that happens only in formal settings such as schools (Philip and Azevedo, 2017). In this content, it will be important to highlight impacts that have benefited underrepresented groups, not just privileged populations. Such impacts can be identified using tools like the Inclusion–Immediacy Criterion framework (Woodson and Boutilier, 2022).

Additionally, research on humor in science communication suggests that there are some situations in which humor can benefit science communication engagement. Yeo and colleagues (2020) found that "respondents who perceived more humor in the video clip (i.e., those in the condition with audience laughter) had more positive views about comedy as a valid source of scientific information" and that the observed relationship was mediated by the perceived expertise of the presenter rather than that person's likability. Another study found that injecting positive humor[123] into a science article increased engagement, especially among non-science majors but that science majors were more likely to be concerned about the credibility of the article when humor was involved (Chan and Udalagama, 2021).

However, there are times when humor can backfire or not be successful, especially when a scientist or engineer is correcting misinformation (Zhang and Lu, 2022). According to this research, *expectancy violation theory* predicts that when scientists or engineers violate expectations of how they might act in a particular context that feels inappropriate, humor will be less well received (for example, making a joke while correcting information about COVID-19 vaccines) (Zhang and Lu, 2022).

This content also has participatory and interactive elements. Although most science communication employs one-way deficit model approaches (Simis et al., 2016), communication scholarship reminds us that people are not passive receivers of messages (Wynne, 1992). New venues for user-generated content such as TikTok allow modally for more audience-centered, interactive, and effective messaging (Lewenstein and Baram-Tsabari, 2022). The importance of

---

[123] "Positive humor" is defined as a funny story, funny comment, joke, professional humor, pun, or cartoon and riddle.

co-creating content with audiences is underscored by that fact that greater involvement is more likely to lead to attitudinal or behavioral change (Villar, 2021).

Using a short-form video platform like TikTok as a channel for this particular audience is also an intentional choice. Most people, especially younger people, get their science information from social media, whether they were seeking it or not (Funk et al., 2017). Engineering and science communication is a part of culture (Davies and Horst, 2016) and can be considered a subset of popular culture; it can even be thought of as a type of fandom (Gartley, 2022). NSF and engineering need to be faithful to their principles to be perceived as an authentic and relevant part of that conversation.

Example 5 References

Chan, A., and C. Udalagama. 2021. Exploring the use of positive humour as a tool in science communication: Do science and non-science undergraduates differ in their receptiveness to humour in popular science articles? *Journal of Science Communication* 20(4):A06.

Davies, S. R. and M. Horst. 2016. *Science communication: Culture, identity and citizenship.* London, New York and Shanghai: Palgrave Macmillan.

Druckman, J. N. 2001. The implications of framing effects for citizen competence. *Political Behavior* 23(3):225–256.

Entman, R. M. 1993. Framing: Toward clarification of a fractured paradigm. *Journal of Communication* 43(4): 51–58. https://doi .org/10.1111/j.1460-2466.1993.tb01304.x.

Funk, C., J. Gottfried, and A. Mitchell. 2017. *Science news and information today.* Pew Research Center, September 20. https://www.journalism.org/wp-content/uploads/sites/8/2017/09/PJ_2017.09.20_Science-and-News_FINAL.pdf (accessed April 23, 2024).

Gartley, L.-E. 2022. CLADISTICS ruined my life: Intersections of fandom, internet memes, and public engagement with science. *Journal of Science Communication* 21(05):Y01. https://doi.org/10.22323/2.21050401

Innovative Research Group. 2017. *Public perceptions of engineers and engineering.* https://engineerscanada.ca/sites/default/files/public-perceptions-of-engineers-and-engineering.pdf (accessed March 9, 2023).

Lewenstein, B., and A. Baram-Tsabari. 2022. How should we organize science communication trainings to achieve competencies? *International Journal of Science Education, Part B* 12(4):289–308.

NAE (National Academy of Engineering). 2008. *Changing the conversation: Messages for improving public understanding of engineering.* Washington, DC: The National Academies Press. https:// doi.org/10.17226/12187.

Philip, T. M., and F. S. Azevedo. 2017. Everyday science learning and equity: Mapping the contested terrain. *Science Education* 101(4):526–532.

Simis, M. J., H. Madden, M. A. Cacciatore, and S. K. Yeo. 2016. The lure of rationality: Why does the deficit model persist in science communication? *Public Understanding of Science* 25(4):400–414.

Villar, M. E. 2021. Community engagement and co-creation of strategic health and environmental communication: Collaborative storytelling and game-building. *Journal of Science Communication* 20(1):C08.

Woodson, T., and S. Boutilier. 2022. Impacts for whom? Assessing inequalities in NSF-funded broader impacts using the Inclusion-Immediacy Criterion. *Science and Public Policy* 49(2):168–178.

Wynne, B. 1992. Misunderstood misunderstanding: Social identities and public uptake of science. *Public Understanding of Science* 1(3):281–304.

Yeo, S. K., A. A. Anderson, A. B. Becker, and M. A. Cacciatore. 2020. Scientists as comedians: The effects of humor on perceptions of scientists and scientific messages. *Public Understanding of Science* 29(4):408–418.

Zhang, A. L., and H. Lu. 2022. No laughing matter: Exploring the effects of scientists' humor use on Twitter and the moderating role of superiority. *Science Communication* 44(4):418–445.

Appendix C
Biographic Sketches of Committee Members and Project Staff

COMMITTEE MEMBERS

Dan E. Arvizu, Ph.D. (NAE; Chair), was formerly the chancellor of the New Mexico State University System and a professor in their Department of Mechanical and Aerospace Engineering. He took the positions after serving as director and chief executive of the National Renewable Energy Laboratory, serving as the first Hispanic to lead a Department of Energy national lab. Prior to that, Dr. Arvizu was group vice president, energy and environment, and systems, at CH2M HILL. He also held executive positions in energy, materials, and technology transfer at the Sandia National Laboratories and began his career as a member of the technical staff at the AT&T Bell Labs. Dr. Arvizu has extensive experience in materials science applications for nuclear weapons and energy systems and development of renewable energy systems, including solar thermal, photovoltaic, and concentrating solar collectors. Among many honors, he received the 1996 Hispanic Engineer's National Achievement Award for Executive Excellence and was inducted into the Great Minds in STEM Hall of Fame. In 2004 Dr. Arvizu was appointed by President George W. Bush, and subsequently in 2010 reappointed by President Barack Obama, to serve 6-year terms on the National Science Board (NSB), the governing body of the National Science Foundation. He was twice elected NSB chairman by his peers, becoming the first Hispanic to hold that position. He presently serves on President Biden's Council of Advisors for Science and Technology (PCAST). Dr. Arvizu earned his B.S. from New Mexico State University and his M.S. and Ph.D. from Stanford University, all in mechanical engineering. He was elected a member of the National Academy of Engineering in 2014 "[f]or leadership in the renewable and clean energy sectors, and for promoting national balanced energy policies."

Edward H. Frank, Ph.D. (NAE), is co-founder and chief executive officer of Brilliant Lime, Inc., and Cloud Parity, both social/mobile software firms. Previously he was a vice president at Apple, Inc., and corporate vice president research and development at Broadcom. Earlier, Dr. Frank co-founded and led the engineering group for Broadcom's Wireless LAN business, which is now one of the company's largest business units. He joined Broadcom in 1999 following its acquisition of Epigram, Inc., where he was the founding chief executive officer and executive vice president. From 1993 to 1996, he was a co-founder and vice president of Engineering of NeTpower Inc., a computer workstation manufacturer. From 1988 to 1993, Dr. Frank was a distinguished engineer at Sun Microsystems, Inc., where he co-architected several generations of Sun's SPARCstations and was a principal member of Sun's Green Project, which developed the precursor to the Java cross-platform web programming language. He holds over 40 issued patents. Dr. Frank is a university life trustee of Carnegie Mellon University and a member of its board's executive committee. He earned a B.S. and M.S. in electrical engineering from Stanford University and a Ph.D. in computer science from Carnegie Mellon University. Dr. Frank was elected a member of the National Academy of Engineering in 2018 "[f]or contributions to the development and commercialization of wireless networking products."

Selda Gunsel, Ph.D. (NAE), is president of Shell Global Solutions (U.S.) and vice president of Global Lubricants and Fuels Technology for Shell. She has served in a number of roles for that company including vice president of fuels and engine vehicle technology, general manager of global products and quality, general manager of lubricants technology Americas, and general manager of Global Strategic Research and Development. Earlier, Dr. Gunsel was vice president for technology development and innovation at Pennzoil. While working as a research scientist there, she undertook sabbatical assignments at Imperial College London, publishing papers on antiwear and viscosity modifier lubricant additives. Her technical background is in the multidisciplinary field of tribology (friction, wear and lubrication) including fundamental research, product development and application, technology management, and leadership. Dr. Gunsel's research areas include lubricant base oil technologies, performance-enhancing additives, synthetic hydrocarbons including gas to liquid technologies, materials, durability, and energy efficiency. She has served as the pPresident of the Society of Tribologists and Lubrication Engineers, is a fellow of that society, and is the recipient of its International Award. Dr. Gunsel received her B.Sc. in chemical engineering from Istanbul Technical University and an M.Sc. and Ph.D. in chemical engineering from Pennsylvania State University. She was elected a member of the National Academy of Engineering in 2017 "[f]or leadership in developing and manufacturing advanced fuels and lubricants to meet growing global energy demand while reducing CO_2 emissions."

William S. Hammack, Ph.D. (NAE), is the William H. and Janet G. Lycan Professor of Chemical and Biomolecular Engineering at the University of Illinois Urbana-Champaign. He is the creator and host of the popular YouTube channel "engineerguy" and has recorded numerous radio segments that describe what, why, and how engineers do what they do for such outlets as American Public Media's *Marketplace* and Radio National Australia. Dr. Hammack's outreach work has been recognized by the National Association of Science Writer's Science in Society Award; the American Chemical Society's Grady–Stack Medal, and the American Institute of Physics' Science Writing Award. In 2021 he was awarded the National Science Board's Public Service Award, which is granted to individuals and groups that have contributed substantially to increasing public understanding of science and engineering. Dr. Hammack's books include *Why Engineers Need to Grow a Long Tail: A Primer on Using New Media to Inform the Public and to Create the Next Generation of Innovative Engineers*; and *How Engineers Create the World*. Each semester, his course "The Hidden World of Engineering" is offered to a diverse mix of students majoring in commerce, architecture, photography, history, and graphic arts and gives students an appreciation for how engineers think. He earned a B.S. in chemical engineering from Michigan Technological University and M.S. and Ph.D. degrees at the University of Illinois. Dr. Hammack was elected a member of the National Academy of Engineering in 2022 "[f]or innovations in multidisciplinary engineering education, outreach, and service to the profession through development and communication of internet-delivered content."

Eboney Hearn, Ed.M., is the executive director of the Office of Engineering Outreach Programs (OEOP) at the Massachusetts Institute of Technology (MIT). In that capacity she oversees the strategic implementation of outreach programs offered through the MIT School of Engineering, focusing on bringing underrepresented and underserved students to the engineering and science fields. Prior to joining the OEOP, Ms. Hearn served as assistant dean for graduate education, diversity initiatives at the MIT Office of Graduate Education. Earlier, she was program director of the Diversity Initiative at the Eli and Edythe L. Broad Institute of MIT and Harvard. Prior to coming to MIT, Ms. Hearn taught mathematics at public middle- and high schools in Boston for 5 years. Before that, she was a manufacturing engineer at IBM, where she led several manufacturing processes in circuit board printing and co-patented a novel photolithography process. Ms. Hearn is a member of the MIT Diversity Think Tank, the Diversity Advisory Committee of the Keystone Symposia on Molecular and Cellular Biology, and the Steering Committee of the UMass Amherst Researchers, Educators, and Business Leaders Network. She holds an undergraduate degree in chemical engineering from MIT, and an Ed.M. from Harvard University.

Laura A. Lindenfeld, Ph.D., is the dean of the School of Communication and Journalism and vice provost for academic strategy and planning as well as the executive director of the Alan Alda Center for Communicating Science at Stony Brook University. As the Alda Center director, she oversees an organization that has trained over 18,000 scientists worldwide and introduced over 50,000 people to the Alda Method. The center provides international leadership in conducting and connecting research and practice to advance clear science and medical communication. She has sought to help people understand how they can support effective stakeholder engagement, build strong interdisciplinary teams, and communicate science more effectively. In her capacity as dean, she oversees a school that prepares undergraduate and graduate students for dynamic careers in media industries. Dr. Lindenfeld's research has been supported by the National Science Foundation, the National Oceanic and Atmospheric Administration Sea Grant Program, the U.S. Department of Education, and the U.S. Department of Food and Agriculture. She has published over 50 peer-reviewed articles, chapters, and reviews in a range of journals including *Science Communication, Environmental Communication*, and *Sustainability Science, Communication and Critical/Cultural Studies*. Dr. Lindenfeld holds an M.A. from the Rheinische Friedrich-Wilhelms-Universität Bonn and a Ph.D. in cultural studies from the University of California, Davis.

Theresa A. Maldonado, Ph.D., PE, is the systemwide vice president for research and innovation at the University of California Office of the President. Previously, she served as dean of engineering at The University of Texas at El Paso. Dr. Maldonado's academic career spans 31 years, including research administration appointments at four other universities: The University of Texas Rio Grande Valley, Texas A&M University, Texas A&M Health Science Center, and The University of Texas at Arlington. She also served in system-level roles as associate vice chancellor for research at the Texas A&M University System, as deputy director of the Texas A&M Engineering Experiment Station, and as founding director of the Texas A&M Energy Institute. Dr. Maldonado has extensive experience at the federal level in advancing engineering research, education, and commercialization initiatives. Beginning in January 2011, she served for 4 years as a division director in the Engineering Directorate at the National Science Foundation (NSF). Her initial appointment at NSF was in 1999, when she served 2 years as a program director in the Engineering Research Centers program and represented the Engineering Directorate on several NSF-wide committees, including the CAREER and ADVANCE programs. Before entering academia, Dr. Maldonado was a member of technical staff at AT&T Bell Laboratories. She earned the B.E.E. with highest honors, M.S.E.E., and Ph.D. degrees in electrical engineering from the Georgia Institute of Technology. Dr. Maldonado is a registered professional engineer in Texas.

Louis A. Martin-Vega, Ph.D., (NAE) is the former dean of engineering, and a distinguished university professor of industrial and systems engineering at North Carolina State University. He joined NC State in 2006 after serving as dean of engineering at the University of South Florida. His academic career spans 34 years and also includes administrative and academic appointments at Lehigh University, Florida Institute of Technology, University of Florida, and University of Puerto Rico at Mayagüez. Dr. Martin-Vega's research and teaching interests are in production and manufacturing systems, logistics and distribution, operations management, engineering education, and broadening participation in the field of engineering. Dr. Martin-Vega has also held several prestigious positions at the National Science Foundation (NSF), including acting head of its Engineering Directorate and director of its Division of Design, Manufacture, and Industrial Innovation. His efforts at NSF led to the creation of the foundation-wide Grant Opportunities for Academic Liaison with Industry (GOALI) program as well as many other initiatives that enhanced industry–university research collaboration and greater inclusion of women and underrepresented minorities in engineering education and research. He is a fellow of the American Association for the Advancement of Sciences, the Institute of Industrial and Systems Engineers, and the Society of Manufacturing Engineers. Dr. Martin-Vega earned his B.S. in industrial engineering from the University of Puerto Rico at Mayagüez, an M.S. in operations research from New York University, and M.E. and Ph.D. degrees in industrial and systems engineering from the University of Florida. He was elected a member of the National Academy of Engineering in 2021 "[f]or support of engineering and engineering education through industry–academic collaboration and opportunities for underrepresented groups."

Yu Tao, Ph.D., is an associate professor of sociology at the Stevens Institute of Technology. In her research, she analyzes issues related to human resources in science, technology, engineering, and mathematics (STEM) as well as online privacy literacy from the sociological perspective. Dr. Tao also investigates how the general public's online privacy skills are affected by their demographic characteristics, online experience, and online privacy educational tools. She is a co-editor of the book *Changing the Face of Engineering: The African American Experience*, published by the Johns Hopkins University Press in 2015, and her research has appeared in social science journals, including *Sociological Spectrum*, *Minerva*, *American Behavioral Scientist*, *Journal of Women and Minorities in Science and Engineering*, and *International Journal of Gender, Science and Technology*. Dr. Tao has served as a co-principal investigator of three National Science Foundation grants and a consultant and co-organizer of a workshop sponsored by the Alfred P. Sloan Foundation addressing her research topics. She received her B.A. in English from East China Normal University, Ed.M. in educational media and technology from Boston University, and M.S. and Ph.D. in the sociology of science and technology from the Georgia Institute of Technology.

Jimmy Williams Jr., Ph.D., is a senior vice president and chief technology officer of ATI, Inc., a high-performance materials manufacturer. He was formerly a Distinguished Service Professor of Engineering and Public Policy and the director of the Engineering and Technology Innovation Management Program at Carnegie Mellon University. From 2012 to 2015, Dr. Williams held the position of vice president of global engineering at Pall Corp, where he led a 750-member engineering unit, driving Pall's global growth initiatives across its life science and industrial products business. Prior to his position at Pall, he spent 10 years with Alcoa, Inc. In his role as senior director of research and development at Alcoa Technology Advantage, he led all facets of business and technology management. As a product innovator, Dr. Williams spearheaded a critical assessment of Alcoa's aero-structures business and led a team that developed an innovative cosmetic finish for Apple's Macintosh computer. Beginning in 1983, Dr. Williams led a nearly 20-year distinguished career at The Boeing Co. where he held a number of significant research and development and program management positions. Among his accomplishments there, Dr. Williams was named Boeing's Black Engineer of the Year in 2001. He earned a B.S. in mechanical engineering from Texas A&M University, an M.B.A. in marketing and management from Lindenwood College, and a Ph.D. in engineering and policy from Washington University.

Jeffrey R. Yost, Ph.D. is the director of the Charles Babbage Institute for Computing, Information, and Culture and a research professor in the history of science, technology, and medicine at the University of Minnesota, Minneapolis. His primary areas of research are the business, social, and cultural and intellectual history of computing, and he is a science and technology oral history specialist. Dr. Yost also has a deep interest in political economy, societal implications of computing structuring and use, and the history of cognitive science and human–computer interaction. His current work includes a monograph on the history of computer security. On the editorial front, he is a past editor-in-chief of *IEEE Annals of the History of Computing* and is on the editorial board of the *Annals* as well as of the journal *Information & Culture*. He also is co-editor of the *Studies in Computing and Culture* book series for Johns Hopkins University Press. Dr. Yost is the co-author of *FastLane: Managing Science in the Internet World*, a book on the National Science Foundation's grant-management system that assesses its impact on cutting-edge scientific research. He earned a B.A. in history from Macalester College, an M.A. and Ph.D. in the history of technology and science from Case Western Reserve University, and an M.B.A. from the Carlson School of Management of the University of Minnesota, Minneapolis.

PROJECT STAFF

David A. Butler, Ph.D., is the J. Herbert Hollomon Scholar of the National Academy of Engineering (NAE). He also serves as the director of NAE's Cultural, Ethical, Social, and Environmental Responsibility in Engineering program. Before joining the National Academies, Dr. Butler served as an analyst for the U.S. Congress Office of Technology Assessment, was a research associate in the Department of Environmental Health of the Harvard T.H. Chan School of Public Health, conducted research at Harvard's John F. Kennedy School of Government, and practiced as a product safety engineer at Xerox Corporation. He has directed numerous National Academies studies on environmental health and technology policy topics, including ones that produced the reports *Climate Change, the Indoor Environment, and Health*; *Damp Indoor*

Spaces and Health; and *Clearing the Air: Asthma and Indoor Air Exposures*. Dr. Butler earned his B.S. and M.S. degrees in electrical engineering from the University of Rochester and his Ph.D. in public policy analysis from Carnegie Mellon University. He is a recipient of the National Academies' Cecil Medal for Research.

Courtney Hill, Ph.D., was formerly a program officer at the National Academy of Engineering working within the Cultural, Ethical, Social, and Environmental Responsibility in Engineering Program. Prior to joining the National Academy of Engineering, Dr. Hill was a Mirzayan Science and Technology Policy Fellow at the InterAcademy Partnership, where she coordinated international meetings addressing how academies across the globe could work together to support the United Nations' Sustainable Development Goals. In addition, Dr. Hill has also taught English at a magnet high school in South Korea as a Fulbright Scholar. Dr. Hill earned her B.S. degree in civil engineering from the University of Arkansas and her M.E. and Ph.D. degrees in civil engineering from the University Virginia. Her doctoral research investigated the relationship between human health and access to silver-embedded ceramics as well as other mechanisms by which silver can be used to treat water in low-income areas.

Maiya Spell, B.S., was formerly a senior program assistant in the Program Office of the National Academy of Engineering. Ms. Spell graduated from the University of Maryland, College Park, in 2021, where she received a B.S. in public health science and certificate in Black women's studies. During her undergraduate career she worked across a variety of fields, including interning in the cardiology department at the University of Maryland Medical Center; interning at Time Organization Inc., a mental health clinic for kids and adolescents; and working in property management for Morgan Properties.

Casey Gibson, M.S. E.I.T, is an associate program officer at the National Academy of Engineering where she focuses on projects related to cultural, ethical, social, and environmental responsibility. In 2022 Ms. Gibson earned her M.S. degree in humanitarian engineering and science with a focus in environmental engineering from the Colorado School of Mines. During her master's degree work, she developed, taught, and implemented a participatory methodology for sociotechnical analysis in engineering projects and focused her fieldwork in rural Colombian communities. She holds dual undergraduate degrees in biological/agricultural engineering and Spanish with a minor in sustainability from the University of Arkansas. Additionally, she was a Fulbright Scholar in Mexico from 2018 to 2020.

Chessie Briggs, B.A., is a senior program assistant in the Program Office of the National Academy of Engineering. Ms. Briggs graduated from the University of Redlands in 2022, where she received a B.A. in both public policy and political science. During her undergraduate career she worked for an international nonprofit organization, traveling to China to help implement a new program in a local orphanage and worked for the (Washington State) City of Federal Way's economic development director, assessing the city's capabilities to host a large-scale event. Additionally, she has recently worked as a legislative intern on Capitol Hill.

Sudhir K. Shenoy, M.S., Ph.D., is an associate program officer in the Program Office of the National Academy of Engineering where he focuses on projects related to practices for engineering education and research. Dr. Shenoy earned his M.S. and Ph.D. in computer engineering from the University of Virginia (UVA), with a research focus on artificial intelligence and robotics and, in particular, the development of emotion adaptive social robots. During his master's degree work, he received training in science and public policy and in engineering ethics at UVA. He developed and taught various robotics and engineering ethics courses for both undergraduate and graduate levels. Dr. Shenoy holds a bachelor's degree in electrical engineering from Jain University in India. He was a UVA–MIT engineering ethics and policy intern at the National Academy of Engineering in 2019.